P9-CLB-383

FLIGHT 427

FLIGHT 427

Anatomy of an Air Disaster

Gerry Byrne

C

COPERNICUS BOOKS

An Imprint of Springer-Verlag

Published in the United States by Copernicus Books,
an imprint of Springer-Verlag New York, Inc.
A member of BertelsmannSpringer Science+Business Media GmbH

Copernicus Books
37 East 7th Street
New York, NY 10003
www.copernicusbooks.com

Library of Congress Cataloging-in-Publication Data
Byrne, Gerry.
Flight 427 : anatomy of an air disaster / Gerry Byrne.
p. cm.
Includes bibliographical references and index.
ISBN 0-387-95256-X (alk. paper)
1. USAir Flight 427 Crash, 1994. 2. Aircraft accidents—Pennsylvania—Hopewell (Beaver County : Township). 3. Aircraft accidents—Investigation—United States. I. Title.
TL553.525.P4 B97 2002
363.12'465—dc21 2002022483

Manufactured in the United States of America.
Printed on acid-free paper.

9 8 7 6 5 4 3 2 1

ISBN 0-387-95256-X SPIN 10796069

For Marie, my wife, who had faith in me when I didn't

PREFACE

This book tells the story of the 1994 crash of a commercial jetliner on its final approach to Pittsburgh International Airport. The disaster that struck the plane, USAir Flight 427, killed all 132 passengers and crew on board, and as with all such tragic crashes, this one created a momentary surge of public grief and a brief rush of interest from the media. But it was not the type of crash to draw the intense, sustained fascination of Pan Am Flight 103, which exploded over Lockerbie, Scotland, in 1988, or TWA Flight 800, which plunged into the sea off the coast of Long Island in 1996. However tragic an event it was, to most of the public it shortly became just an "ordinary," not particularly memorable, disaster.

But of course there was almost nothing ordinary about it. This is primarily the story of the dozens of investigators who sorted through the wreckage, and the scientists and engineers who ultimately solved the puzzle at the heart of this and perhaps several other disasters. And there is a wrinkle to the story: one of the "characters," perhaps the central character, is not a human at all, but an airplane.

Until the National Transportation Safety Board (NTSB) delivered its official verdict on the accident in March of 1999, its cause had been one of US aviation's longest-running puzzles. This book explains why it took so long to get to the bottom of this mystery.

Beginning with the investigation of an earlier crash, United Airlines Flight 585, this books shows how only after almost a decade of probing did the brilliant sleuthing and informed hunches of the "tin kickers" at the NTSB pay off—uncovering a rare but nonetheless fatal flaw in the design of the Boeing 737. This plane, the "work horse" of modern commercial airlines, is indisputably the most popular and arguably the safest of all major airliners. Yet the difficulty of getting to the bottom of the 737's problem, and the sometimes competing and sometimes overlapping interests of its manufacturer, the airlines, the pilots' union, the families of those killed, and most of all two agencies of the US government—the Federal Aviation Administration and the National Transportation and Safety Board—come together here in what, after all is said and done, is a kind of engineering detective story.

Gerry Byrne
Dublin, Ireland
January 2002

ACKNOWLEDGMENTS

Although I have only ever been unfortunate enough to be present at the immediate aftermath of one fatal airplane crash, it made an everlasting impression on me. When emergency services allowed my photographer and me into the woods where the four-seater had crashed, killing all aboard, I discovered that I had not long before interviewed its owner-pilot, who had built the aircraft in his own garage. But even this sad, searing memory fades into insignificance with the haunting recollections of one Washington evening, listening to the stories being told by men and women whose lives had been dramatically changed by air accidents. There were airline crash survivors like Tom Eilers, who crawled dazed, almost losing his will to live, out of the burning wreck of a DC-10 at Sioux City, from which 111 less fortunate souls like Tom O'Mara's daughter never emerged. The heartbreaking stories of people like Carole Rietz, who lost a wonderful son in the ValuJet Everglades crash of 1996, graphically illustrated the horrific impact of an airliner crash on relatives of victims, people who often carry mixed burdens of grief, anger, depression, and even pointless guilt for years afterwards.

This book is not about their suffering—it is about discovering why airplanes crash—but their experiences and their courage have deeply inspired me on this journey into the wonderful world of flight, and the bleak world of airplane accidents. Apart from inspiration, no writer can cope without good sources, and I must give thanks to as many as I can. Although not direct sources for this book, a few deserve special mention. Fellow journalist Tom Clark, who first pointed me at this story; Gail Dunham, whose ex-husband, Captain Harold Green, died in the crash of Flight 585 at Colorado Springs, and who goaded me on; the editors at *New Scientist*, who published my initial efforts; Dan Loughrey and Willie Walsh, who squeezed me between the cracks into a Boeing 737 simulator so I could learn how 737s were flown instead of eternally marveling at the apparent wizardry of it all; the various Aer Lingus and Ryanair captains, who, over the years, tolerated me in their cockpits to watch them fly; and the air traffic controllers at Dublin, Ireland, who occasionally suffered me at their elbows.

An immense quantity of research material was provided by the National Transportation Safety Board, in Washington, D.C., which also provided me with access to many of its key personnel for interviews. They included Tom Hauteur, investigator in charge (Flight 427); Al Dickinson, investigator in charge

(Flight 585); Greg Phillips, Systems Group chairman; Malcolm Brenner, Human Factors Group chairman; Cynthia Keegan, Structures Group chairman; Frank Hildrup, Structures Group member; Al Lebo, Air Traffic Control Group chairman; Greg Salotollo, Meteorological Group chairman (Flight 585); Tom Jacky, Aircraft Performance Group chairman (Flight 427) and Flight Data Recorder Group chairman (Flight 585); Dennis Crider, kinematics, and Jim Cash, acoustic expert. Although they did no work on either of the main crashes featured in this book, Jim Hookey (Powerplants) and Nora Marshall (Survival Factors) kindly and patiently explained how their colleagues went about their work. Jamie Finch and Sharon Bryson described the work of the Family Affairs Office, and Ted Lopatkiewicz from Public Affairs coordinated things splendidly.

Former NTSB chairman Jim Hall came out of retirement, as it were, to discuss Flight 427 with me, as did former NTSB board member Bob Francis, in addition to Bernie Loeb, former head of the Office of Aviation Safety; Ron Schleede, former deputy director of the Office of Aviation Safety; and Chuck Leonard, who had led the first Operations Group on Flight 427. Some other former NTSB and Boeing staffers spoke to me but requested that I preserve their anonymity.

Many perplexing questions about Rocky Mountain weather were answered by Al Bedard of the National Oceanic and Atmospheric Administration, Terry Clark of the National Center for Atmospheric Research, and Professor John Marwitz, formerly of the Atmospheric Science Department of the University of Wyoming at Laramie. Paul Knerr of Canyon Engineering led me through the topic of silting in airplane hydraulics, and Michael Demetrio of Corboy & Demetrio spoke to me about the legal process followed by relatives.

I was aided by several pilots, including Captain John Cox, who was an ALPA delegate to the Systems Group; Captain Robert Sumwalt, an ALPA delegate to the Human Factors Group; and Captain Joe Kohler, who represented ALPA on the Engineering Test and Evaluation Board. Dave King of the United Kingdom Air Accident Investigation Branch provided an interesting foreign perspective on the investigation.

I leaned heavily on the work of some other authors, especially Macarthur Job, whose excellent and brilliantly illustrated series *Air Disasters* (Volumes 1–3) is the closest thing there is to an encyclopedia of airliner crashes, and which was almost always open on my desk. I also consulted Christine Negroni's *Deadly Departure* (HarperCollins, 2000) about TWA 800, and Stephen Frederick's *Unheeded Warning* (McGraw-Hill, 1996) for the background to the Roselawn, Indiana, crash. Andrew Weir's *The Tombstone Imperative* (Simon & Schuster, 1999) provided illuminating analysis of a wide range of air disasters. Also consulted were Nicholas Faith's *Black Box* (Boxtree, 1996), Stanley Stewart's *Emergency: Crisis on the Flight Deck* (Airlife, 1999), and *Aircraft Accident*

Analysis: Final Reports (McGraw-Hill, 2000) by James Walters and Robert Sumwalt. David Beaty's *The Naked Pilot* (Airlife, 1999) proved a provocative introduction to the topic of human factors in aviation accidents.

Robbie Shaw's *Boeing 737-300 to 800* (Airlife, 1999) was a useful source on the technology and history of the Boeing 737; other historical perspectives were provided by T. A. Heppenheimer's *Turbulent Skies* (John Wiley & Sons, 1995) and Matthew Lynn's *Birds of Prey* (Heinemann, 1995). I found illumination on many questions of aircraft performance in Barnard & Philpott's revision of Kermode's classic, *Mechanics of Flight* (Longman, 2000).

Inevitably I was drawn to the NTSB's own reports into accidents, especially on Flight 427 and Flight 585, in addition to the report on the Exxon Valdez disaster and others I have analyzed over the years. These take two forms: factual reports by individual groups (e. g., Systems), and the final report drawn together by the investigator in charge and passed by the board of the NTSB. Factual reports, and the appendices and correspondence they attract, are a vital source of information for the serious student of airplane crash investigation. I also mined the FAA's *Critical Design Review of the Boeing 737* and the report of the *Engineering Test and Evaluation Board* (FAA, 2001). The Rand Institute for Civil Justice's study on the NTSB, *Safety in the Skies* (1999), provided an interesting perspective on the Party System of air crash investigation. Reports in many US newspapers and magazines provided invaluable signposts to this story, but nowhere was this story covered with such regularity and depth as in the *Seattle Times*.

Although Boeing had generously cooperated with my researches on other projects, for this one the plane maker adopted a policy of *omertà* and, not for the first time in connection with Flight 427, refused to sanction interviews by writers with its staff. Readers may draw their own conclusions from this—I offer no suggestions because I have none. However, reconstructing Boeing's participation in this investigation was relatively straightforward from the comments of other parties and because of the large number of reports and letters its staff wrote, made available to me from the NTSB's archives.

Finally, a special word of thanks to Paul Farrell, Editor-in-Chief at Copernicus Books, and his hard-working colleagues and freelancers: Mareike Paessler, Jordan Rosenblum, Tim Yohn, and Lyman Lyons.

INTRODUCTION

When the Wright Brothers first took to the air in *Flyer* at Kitty Hawk in 1903, they crashed on their fourth flight, setting an unwelcome ratio of accidents to flights for aviation's first year. Twenty-eight years later, when a group of 24 pilots gathered to form the first pilots' union, air safety had improved dramatically, yet by today's standards it was still appalling; half of those men were destined to die in airliner crashes. It was not until 1954 that air crash casualties were first outnumbered by those dying in train wrecks. In that year, three US airliners crashed and 90 people died. Were the same proportion of the traveling American public to die today, there would be at least 20 major air disasters annually in US airspace, with a death toll in the thousands. Although it is almost unforgivable to attempt to draw comfort from the year 2001, which was blighted by four horrific crashes killing 391 passengers and crew plus thousands on the ground as a result of terrorist action—yet there was only one major jetliner crash that year due to what might be called accidental causes: the November 12 crash of an American Airlines Airbus in Queens, New York, killing all 275 aboard.

There were two reasons for a gradual, if sometimes erratic, improvement in air accident statistics. First, a legion of rules and regulations about flying were promulgated by Government, initially by the US Department of Commerce, then the Civil Aeronautics Board and its successor, the Federal Aviation Administration. New airplane designs had to be government-approved. Pilots had to be better trained. They received more accurate weather forecasts. And a nationwide system of air traffic controllers, radar, and radio beacons replaced a haphazard system of lights atop towers and stopped pilots getting lost or flying blindly into each other.

A second reason for the reduction in air accidents was the work of air accident investigators, the people whose work on major air accidents fills the pages of this book. Over 75 years, they probed crashes to find out what happened, not so much to allocate blame, but to make sure the same thing never happened again. At first, following the passage of the 1926 Air Commerce Act, they operated as a division of the US Department of Commerce, but later worked within the Civil Aeronautics Board's Bureau of Aviation Safety. The Civil Aeronautics Board evolved into the Federal Aviation Administration (FAA), but then, when it became obvious that government services themselves might be part of the problem in an air disaster, investigators demanded and secured more autonomy. Congress established the National Transportation Safety Board in 1967 as an autonomous government agency. Its independence was further underlined in 1974, when it was removed from the overall control of the US Department of Transportation. Since its creation, the NTSB has investigated more than 115,000 aviation accidents and incidents, mostly related to light aircraft, and issued many thousands of recommendations for the improvement of aviation safety.

Although it has powers to subpoena witness and documents, the NTSB cannot make new rules or force airlines or plane makers to change their ways. That role is reserved for the FAA. Apart from its findings of the cause of air accidents, the NSTB can only issue recommendations, mostly to plane makers, to airlines, airport authorities, and, finally, to the FAA. Nevertheless, the NTSB likes to boast that up to 84 percent of its recommendations result in the creation of new rules or the correction of unsafe procedures. But a worrying 16 percent, at least, are ignored or rejected. We now know that at least one major airborne disaster could have been averted if more NTSB recommendations were heeded. After a fire in the cargo hold of an American Airlines DC-9 that landed safely at Nashville in February 1988, the NTSB recommended that smoke detectors, among other fire safety equipment, be installed in the cargo holds of all aircraft. The call went unheeded by the FAA and by the airlines and plane makers, some of whom actively lobbied against the proposal. On May 11, 1996, 110 died when a ValuJet DC-9 crashed in flames into the Florida Everglades while attempting to return to Miami after the crew had discovered a fire raging in the cargo hold shortly after takeoff. A smoke detector could have alerted them earlier, perhaps even before takeoff, and saved many lives. Two years later, the FAA finally mandated the installation of smoke detectors in cargo holds but gave airlines up to three years of grace before fitting them.

This book traces the investigation by the NTSB into the tragedy of USAir Flight 427, which crashed into a lonely hillside outside Pittsburgh in September 1994, killing all 132 aboard. It was a remarkable accident for several reasons. It seemed to have had no obvious cause, and it struck a Boeing 737, arguably among the safest and easily the most popular airplane ever built. It also mirrored the mysterious crash three and a half years earlier of another 737, this time belonging to United Airlines, with which we commence this story. When United Flight 585 crashed at Colorado Springs one blustery March day in 1991, all 25 aboard died, and the cause was always far from clear. In December 1992, the NTSB admitted it had failed to solve the case of Flight 585, declaring that there was no obvious cause, although it did suggest that either weather or mechanical failure might be to blame. Their inability to solve Flight 585 hung like a dark cloud over the investigators. It had become one of just a handful of crashes for which they had failed to find a cause since the NTSB's foundation 25 years earlier.

When confronted with a second, apparently insoluble accident at Pittsburgh, the bitter memory of Flight 585 goaded investigators, stinging them like an irritating insect and propelling them into investigating that second crash like no other accident they had ever handled before.

In most airliner crashes, good clues usually emerge sooner or later. On the radio to controllers, a pilot may describe an emergency like a fire, an engine failure, loss of control, or atrocious weather conditions. One of the black boxes, the flight data recorder may reveal some anomaly in the airplane's flight behavior. Similarly, the other black box, the cockpit voice recorder, may betray the fact that the pilots are lost, or improperly manipulating the controls, or omitting to follow proper procedures. Should the recorders fail to yield clues, a painstaking search of the wreckage will often yield a fatal flaw. A vital part may have failed due to wear and tear, or lack of maintenance, or improper installation or adjustment. Most crash investigations are well on their way to being wrapped up within a year or 18 months. But Flight 585, and Flight 427 afterward, offered no clues whatsoever. Nothing in communications with the tower, nothing in the black boxes, nothing unusual in the wreckage.

Reconstruction of the weather conditions the day Flight 585 crashed revealed unusual, freakish winds. Boeing proposed convincingly that they could have downed the aircraft. Investigators were not so sure but left the possibility open, along with mechanical failure. A reconstruction of the flight path followed by Flight 427 in its final seconds showed a remarkable similarity with that of Flight 585, yet the weather that day in Pittsburgh had been almost calm. It was only when investigators started looking at reports from Boeing 737 pilots whose aircraft had inexplicably misbehaved that the chilling possibility emerged that all Boeing 737 aircraft might possess a fatal design flaw—a freak blemish which, without warning, could tear an otherwise perfect airplane out of the sky in seconds. To inconclusively uncover this flaw took a total of eight years, from when Flight 585 crashed until the NTSB formally delivered its verdict on Flight 427. It had been the longest, most complicated, and one of the most expensive aviation crash probes ever.

Unlike the FAA, which employs some 50,000 people, the NTSB is a small agency. At the time of these crashes, it employed about 400 people divided between its Washington, D.C., headquarters and a network of local field offices. Not all worked on air crashes. A significant proportion of them were devoted to accidents on roads, rail, commercial shipping and pleasure boats, and even gas and oil pipelines. In the early days, investigators were sometimes expected to handle accidents across a wide range of modalities but, increasingly, they specialized in one mode of transport only. Nevertheless, it was not long since an NTSB psychologist might be asked to investigate the behavior of a ship's captain one day, then probe the background of dead jetliner pilots not long afterward.

Skilled, experienced aerospace engineers have always been in strong demand at the NTSB but, despite a small but steady increase in staffing across the years, investigators always found themselves playing catch-up with aviation technology. While it was possible that a clever man in a field office might understand every control and rivet in a Cessna 172 or a Piper Cub, the complexity of a modern transport category airplane completely rules out that sort of familiarity. In an age of specialization, engineers increasingly attach themselves to different parts and functions of the airplane. One investigator might be an expert on the structure of an airplane, another on controls, yet another devoting him- or herself entirely to engines. Yet no two airplanes designs were alike. For example, at the time of these accidents, there were two US manufacturers of large transport aircraft, Boeing and McDonnell Douglas (they have since been merged). A third, Lockheed, had quit passenger aircraft manufacture but its L-1011 Tri-Star still plied air routes. Of Boeing aircraft, for example, there were six main models and dozens of variants. McDonnell Douglas had concentrated most of its recent production around two models but, again, there was a large number of variants. In Europe, Airbus was producing almost as wide a range of jetliners as Boeing and had pioneered the use of complex, computerized cockpits. In addition, there were smaller jetliners from Fokker, British Aerospace, and Bombardier/DeHavilland, among others.

The picture was just as bewildering at the turboprop end of the commuter aircraft, mainstay of the regional and feeder airlines. Most of those aircraft were coming from French, British, Swedish, Canadian, and even Brazilian manufacturers. Not only was each aircraft make and model built differently, but many models had idiosyncrasies that required unique skills and training to fly them. The Captain of a DC-9 could not possibly hope to step into the cockpit of a Boeing 747 and commence flying it without months of retraining and familiarization. The reverse was also true.

For years, the NTSB had recognized that no one investigator could possibly probe all aspects of a major air crash. Instead, under the leadership of a senior investigator (the investigator-in-charge) it deployed a wide range of specialties needed to tackle different aspects of a crash. There is a structures investigator who determines if there are flaws with the wings, or the fuselage, or perhaps even the undercarriage. This investigator usually takes responsibility for the crash site, ensuring that all the wreckage is accounted for and that proper trails of evidence are followed. Another person takes responsibility for the engines, yet another for the aircraft's controls (or systems), while a fourth investigator probes the airline's operations and the crews' training and suitability for the flight. A staff psychologist might also get involved to see if there are any hidden aspects of the aircraft or the airline's procedures that trip the unwary into making fatal errors. Other specialists take charge of the aircraft's black boxes and decipher their contents. Another investigator reviews the accident flight's relationship with air traffic controllers, while a staff meteorologist analyzes the weather at the time of the crash to determine if it had a bearing on the accident. There are metallurgists, sound specialists, fire and explosions experts, simulation experts, survival experts, maintenance experts, even pilots, all bearing down with their unique skills in bid to uncover the cause of an accident.

Yet many times, this is not enough. The NTSB's experts often need to draw on the knowledge of other investigators, in Britain, in France, sometimes even in the former USSR, to come up with answers to unique problems involving aircraft from those countries, or to help with investigative techniques they have pioneered. British investigators have, for example, long mastered the art of reconstructing wreckage, while Russian specialists have added a new dimension to the interpretation of cockpit voice recorder tapes. US investigators, in turn, are regularly called overseas to help solve crashes involving US-built aircraft. And it is still not enough. The differences between aircraft makes and types is often so great that it might take investigators months, even years, to get fully up to speed.

Rather than collapse under a mountain of paperwork and study each time an airplane crashes, the NTSB resorts to what it calls the Party System. Any organization even remotely likely to be implicated in an air accident may apply to be affiliated to the investigation as a party. They participate in the probe, perhaps even suggesting avenues of further exploration to investigators. In their turn, the investigators draw upon the knowledge and experience of the party members and use them as a shortcut to obtaining further information. The plane maker is an obvious starting point.

When a Boeing aircraft crashes anywhere in the world, a small team of Seattle-based engineers springs into action. Its composition mirrors that of a typical NTSB team with structures, systems, operations, sometimes even data recorder specialists. Depending on the severity of the accident, they assess the skills likely to be in demand at the crash site. The team, in turn, can draw upon engineers from elsewhere in the company, and within hours a group of specialists is en route to the disaster scene, ready to assist NTSB personnel make sense of it all. The airline will also dispatch as full a team as it can muster, although only the Federal Aviation Authority is legally entitled to be represented on every investigation.

Similarly, an engine manufacturer will also dispatch engineers familiar with the crashed engine's powerplants, a hydraulics company will want to be represented on the team that probes the control systems, the undercarriage manufacturer on the Structures Group, and so forth. Labor unions also get involved. The International Association of Aircraft Machinists attaches volunteers to several groups, especially Structures, Powerplants, and Systems. And the Airline Pilots' Association (ALPA) is often among the most represented of all the parties. Next to the FAA, the plane maker, and the airline, ALPA is often the only party with delegates on practically every group in an investigation.

The NTSB gains ready access to immense quantities of knowledge, experience, and expertise as a result of the Party System. The benefits to the parties themselves are less obvious but they can be significant. On a more altruistic level, their inputs may lead to a speedy conclusion of a crash probe, hopefully resulting in safety recommendations that will prevent a similar accident from reoccurring. Equally altruistically, ready access to all the data flowing in from an investigation can enable an airline, a plane maker, or a component manufacturer to take early action to correct any flaws uncovered in its products or processes.

Motives can be interpreted as being less altruistic the further one moves along the chain of benefits. For example, a party may quickly discover if one of their products (or members, in the case of the labor unions) is under suspicion as a probable cause of the accident. If so, they can then present mitigating factors to the investigators or suggest other contributory factors that should also be taken into account, such as poor weather in the event of pilot error, or pilot error in the event of an airplane malfunction, for example. Corporate lawyers can be tipped off if a flood of liability claims is to be expected. (Despite the NTSB's request that information uncovered in an investigation be kept confidential until reports are published, this often occurs.) And finally, faced with a situation where a party is likely to have to shoulder the blame for its products, or its affiliates, the desperate option remains of attempting to blame something, or somebody, else.

The result is often an interesting tension between investigators and party members during an investigation, a tension which, as can be seen in this book, can occasionally develop into outright hostility. Each investigator acts as chairman to his or her group, which may include delegates from the FAA, the airline, the plane maker, the pilots' union, the machinists' union, and a component manufacturer. Investigators are usually diplomatic about the dynamics of their groups and in any event claim that other group members are often more than able to counter any false allegations or misleading scenarios proffered by any one delegate. For example, an unfounded attempt by the airline to blame the crew will be met with a robust rebuttal by the pilots' union, and vice versa. In other words, they will keep each other in check. This process is often very much in evidence during a public hearing, as party representatives cross-examine witnesses. At least one public hearing and possibly also a public board meeting may be held during an investigation.

However, the conclusion of the investigation offers parties a further opportunity to influence the final outcome. First, each party chairman will have wrapped up his or her separate investigation. In their final reports, most of these will be able to say that no errors, flaws, or failures were uncovered as part of their investigation, and their groups have been disbanded. Those groups who do discover any unusual factors, or outright failures of machinery, or errors by personnel, will say so in their final reports, which are provided to the investigator in charge. He or she then invites all of the parties to the investigation to submit their take on the accident, what they think may have caused it, and how such an occurrence may be prevented in future.

If the issues are straightforward and the parties submit probable causes more or less in agreement with each other, the investigator-in-charge is likely to rapidly draft a finding for transmission to his or her superiors who, in turn, may submit it straight to the board of the NTSB for their adoption. Occasionally, however, party submissions offer conflicting, often risible interpretations of the evidence in which case the final process is likely to be slightly more drawn out. Senior NTSB executives will then exhort investigators to check and double-check their conclusions before submitting them to a final, rigorous analysis. A decision by the board of the NTSB is the culmination of an almost judicial process that is very difficult to challenge, even in the courts, so there is tremendous pressure on the staff to get it right. Reputations can be destroyed and fortunes ruined by its findings. A final, highly choreographed public board meeting then takes place at which, with the assistance of the appropriate party chairmen or staff specialists, the investigator-in-charge introduces the significant evidence. The board then adopts the probable cause and publishes its recommendations. Unlike a public hearing, where representatives of the parties may cross-examine witnesses under oath, the parties play no role in the public board meeting, which is unsworn, and where only board members may ask questions.

The investigations and the events covered in this book raised a serious question mark over the safety of the Boeing 737, the most popular airplane ever built. In February 1990, it eclipsed sales of the world's previous best-selling jetliner, the Boeing 727, which had totaled 1,832 copies delivered. By March 1999, more than 3,000 Boeing 737s were in service worldwide, having accumulated 91 million hours in the air and carried almost 6 billion passengers—nearly the equivalent of the entire world's population. Hundreds more are on order, and more than 800 are in the air at any given moment around the world.

Yet at one stage, investigators privately debated calling for the airplane to be grounded, a call that was made more publicly by one aviation attorney. Had it been heeded, the nation's air transport system would have almost ground to a standstill so vital is the 737 to the system. At least one major airline, Southwest, would have gone out of business because it operates no other airplane. The FAA was so concerned with revelations about the 737 that it conducted its own independent but ultimately inconclusive investigation, separate to the NTSB's long-running probe. During these various probes, several alterations were ordered to the airplane's controls, additional maintenance checks were slated, and pilots had to be trained in special recovery techniques unique to the 737 should they lose control of their aircraft. Even after the NTSB finally announced its dramatic conclusions, a further, more intensive probe by the FAA uncovered more flaws in the airplane. Thousands of 737 aircraft flying today remain under a cloud and will eventually have to be recalled for even further special modifications to make them safer.

The extra work needed on the 737 fleet is ironic because, despite its flaws, it has a reasonably good safety record. The NTSB published calculations by an aviation loss adjuster for the ten years ending December 1997, which showed that 43 Boeing 737s had been write-offs due to accidents. That worked out at one 737 crash for every 1,010,101 takeoffs, or a serious crash rate of 0.99 per million flights—better than the Boeing 727 (1.19 crashes per million flights) or the unlucky Fokker series of airplanes, which crashed almost twice as much (2.23 times per million flights). However, the analysis suggested that the McDonnell Douglas competition to the 737 was safer. Its closest competitor in terms of size and vintage, the DC-9 series (including the later MD80 variants) had an accident write-off rate of just 0.86 per million flights, a 13-percent improvement on the 737's record.

However, a Boeing analysis based on the same figures divided up the 737 fleet between older and newer model variants. This analysis suggested that the older, 737-100 and -200 variants (including Flight 585, which crashed in Colorado Springs) were three times more likely to crash than newer -300 (including Flight 427, which crashed at Pittsburgh), -400, and -500 variants. In the Boeing analysis, the older -100 and -200 series aircraft had a crash rate of 1.5 per million flights, while the newer models crashed just 0.5 times for the same number of departures, an enviable record for any aircraft model.

But this is more than just the story of how the problems with one airplane model were tracked down and solved. Throughout, we attempt to illustrate how lessons learned in earlier crashes have a bearing on the solution of more recent disasters. This book also highlights the failure of the Federal Aviation Administration to act decisively on behalf of the traveling public. Throughout the 1980s and into the mid-1990s, the FAA allowed Boeing to produce new variants of the 737 according to outdated certification procedures that harked back to the 1960s when, if had been designing a brand-new airplane, tougher, far safer standards would have been applied. Had those stricter standards been applied to subsequent variants of the 737, the crash of Flight 427 probably would not have occurred and 132 men, women, and children would be alive today. How much of those luckless FAA decisions were part of its own honest strategy, and how many were at the behest of aggressive Boeing lobbyists, may never be known.

Part One

585

Chapter 1 THE UNFRIENDLY SKIES

Imagine you can see the wind.

What most people fondly suppose to be a smooth, broad, steady stream of moving air is, in fact, no such thing. What the TV forecaster blithely shows advancing across the continent in a neat parabola of parallel arrows is, more often than not, a turbulent maelstrom of conflicting currents and eddies. If you could don magic goggles and watch from an aircraft, you would soon see that the features of the landscape below dictate much of this nearly chaotic flow. Close to the ground, the wind can behave just like a river; it dodges around mountains, rushes down valleys, and—if it contains colder, denser air—pours down hillsides. It foams up where it meets obstructions, like isolated bluffs or tall buildings, often leaving a diminishing trail of miniature whirlpools in its wake. When the wind meets a tall mountain range, it rises and then crashes over the crest like a waterfall, leaving the turbulent signs of its passing in the sky for hundreds of miles downstream.

Some winds are generated by depressions, areas of warm, light, moisture-laden air that suck in drier, heavier cold air from many hundreds of miles away. Others have their origin in more local phenomena—a thunderstorm, for example, or air moving in from a cool sea or lake to replace warm air rising over land heated by the sun. Some, like the Santa Ana winds in California, are so predictable you can almost set your watch by them. But by and large, winds are distinguished by their fickleness.

And none are more fickle than the fierce winds that can hurtle downward from large mountains. Pikes Peak is an outcrop of the Rocky Mountains whose brooding, craggy shape rises to more than 7,000 feet above the plain at its foot and dominates the vista to the west of Colorado Springs. Locals talk of weird winds that seem to appear from nowhere, suddenly tearing wildly through their neighborhoods. Old-timers will tell you that the phenomenon is often at its worst when Pikes Peak sprouts a stationary cap of white, fluffy clouds that never blow downwind with other clouds, despite blustery conditions. There may also be other stationary clouds, they say, such as a series of thin, lens-shaped formations at varying heights above and downwind of the peak. Sometimes, but not always, you can spot a wispy trace on the mountain's downwind flanks that can vanish as quickly as it appeared, then mysteriously appear again. Sometimes you'll see nothing at all because the air is too dry for clouds. But don't take that to mean nothing is going to happen.

On the morning of Sunday, March 3, 1991, Fredo Killing of Childe Drive in Colorado Springs stepped outside his house and scanned the sky. An avid and experienced glider pilot, Killing had planned a flight from the Black Forest airfield. But by 9:00 A.M. or so, he was not sure that gliding was a good idea. Earlier, strong winds he'd estimated at 100 miles an hour had ripped tiles weighing 10 pounds apiece from the southwest side of his roof. Now, with the air around him calmer but still gusty, he keenly watched the morning sky, looking anxiously for tell-tale signs of strong turbulence aloft.

True gliders have no engines. They are towed into the air by conventional planes, and when high enough, the pilot releases the towrope. In poor gliding conditions, when the air is still, they simply nose gently downward, disappointed, and circle back to the airfield. But when conditions are right, gliders can soar like eagles, borne aloft on rising columns of warm air, or hitching rides on ascending currents of air generated by the action of wind on mountainous terrain. Occasionally, they reach the height of cruising jetliners, higher than Mount Everest, where survival for long without bottled oxygen is almost impossible.

High in the sky almost directly above him, Killing saw a series of smooth, lens-shaped high-altitude clouds running parallel to the foothills of the Rockies to the west and extending as far north and south as the eye could see. Despite gusty southwest winds, the clouds remained motionless. Closer to the ground, about 1,500 feet up, he noted numerous fleeting wisps of condensation. These were evidence to his trained eye of whirling bundles of rotating air known as *rotors*, for all the world like slow-motion whirlwinds tilted on their sides.

Killing had learned to respect rotors. Once, at an altitude of more than 10,000 feet, his glider had been turned upside down by one. He righted the glider, but the experience had been unnerving. His near miss had come years ago when he was flying close to the mountains, where rotors were more likely to be encountered. But today, for the first time, he was seeing them above his own home. With his twelve years' experience, Killing rated the conditions as unusually severe—even weird—and decided against flying on this strange morning. Better instead to stick around his house and make plans for getting the roof repaired.

About a half hour later, a few miles away, retired policeman Harold Darnell was driving his pick-up to a local flea market. His wife sat beside him in the passenger seat. He had just turned off Fountaine Boulevard onto Grinnell Street when the truck suddenly slowed down, almost stalling. Grinnell Street is relatively flat, but to Darnell it felt as if the pick-up were struggling uphill in too high a gear. Puzzled at this strange force holding him back, Darnell down-shifted two gears to prevent the engine from stalling. Then, as suddenly as it began, the slowing force disappeared. The engine roared and his truck lurched ahead. A block or so farther on, glancing in his rear-view mirror, he noticed a large cloud of smoke billowing into the sky from the general direction of Widefield Park. A house fire, probably in their area, he told his wife. "Could be a neighbor's, even our house," she replied, alarmed enough to suggest they turn back to check. But Darnell, distracted and puzzled over what had possessed his vehicle, wanted to press on to the flea market.

Around the same time, about 9:40 A.M., another resident of the Widefield Park area, Lester Theusen, was en route to the same flea market in his 1981 Buick Regal. He had debated with himself about going because the morning weather forecast had threatened strong, even gale-force winds, and he wondered if many sellers would risk setting up their stalls. As he had listened to the forecast, the air was calm at his home on Cardinal Street. But he could recall times when it was nearly calm at home and extremely windy just a few miles north along Interstate 25. He entered the interstate at Exit 132 and headed north. Shortly after he passed Exit 135, he was traveling at 55 or 60 mph when an intense gust hit his car on the left side and, as he put it, almost blew him off the road. The incident stuck in his mind because it was only a mile or so from where his daughter's car once rolled over in an accident.

About the time that Harold Darnell's pick-up mysteriously slowed down, Detective Pat Crouch of the Colorado Springs Police Department was driving in the opposite direction along Grinnell Street when his wife, Cynthia, remarked that an airplane was "flying funny." Crouch looked and saw a jetliner with United Airlines' colors, heading north, suddenly turn sharply to the right, roll onto its right side, and then head straight down. Crouch got on his police radio and called the police department, warning them that a plane might be about to crash into the Fountain Valley School. He watched the still-vertical plane continue to dive until it was out of sight behind a ridge. There was the sound of an explosion, and a large fireball erupted into the air.

Minutes earlier, John Sjonost, a real estate salesman and former US Air Force master sergeant, was startled by heavy wind gusts rocking his house on Woodstock Street. One strong gust was followed almost immediately by another "extremely heavy, booming gust." Then his phone rang. It was his oldest daughter, and she was nearly out of breath: a huge plane had just crashed into Widefield Park, right beside her apartment building.

Georgia Matteson, a flying instructor, was in a single-engine Cessna 172 with a pupil, Dave Norgren, heading in to land at Colorado Springs Municipal Airport. The air traffic controller on duty, James Rayfield, had just cleared Matteson to land in a northwest direction on runway number 30 when she saw a United Airlines jetliner above her and to the right. It was making a wide lazy turn to come into the wind for a landing on the adjacent runway, number 35. Airports name their runways according to the points of the compass that they most closely approximate. North corresponds to 0° or 360°, east to 90°, south to 180°, and west to 270°. Matteson and her pupil, headed for runway 30, were approaching from the southeast, pointing their plane northwest at 304°. Runway 35, which the United jetliner was heading for, pointed at 349°, almost due north.

Over her radio Matteson heard what sounded like a pleasant, efficient pilot—a woman—in the cockpit of the United Airlines Boeing 737, receiving clearance for landing from Rayfield in the tower. Rayfield was passing on information about the wind speed and direction, snapping off details unintelligible to anybody but an aviator: "Three-two-zero, one-six, gust two-niner." That meant the wind was coming from 320°, roughly northwest, with a mean speed of 16 knots and occasional gusts as strong as 29 knots. Aviation has put men and women in space, but it still uses the archaic terminology of the sea. A knot is 1 nautical mile per hour, about 1.15 miles per hour. (The word "knot" dates from sailing ships and refers to the rate that a rope unreeled while being dragged behind a ship. The carefully spaced knots on the rope were counted while an hourglass emptied.) Rayfield's staccato message, therefore, contained a caution about wind gusts of 33 mph.

The United pilot responded, confirming the runway details and asking if any aircraft had reported a loss or gain of airspeed. Rayfield told her a Boeing 737 had earlier reported a 15-knot loss at 500 feet, a 15-knot gain at 400 feet, and a 20-knot gain at 150 feet. "Sounds adventurous," she replied from the United cockpit. The numbers suggested that turbulence was affecting airspeed, possibly from a mixture of headwinds and tailwinds at different altitudes. Matteson and Norgren had just experienced an extremely bumpy ride. They reckoned that at one point the Cessna had quickly lost between 300 and 500 feet of altitude, although they remained in control. So while it was a picture-perfect day for flying, with the vast Rockies a stunning backdrop to Colorado Springs and the airport ahead, there was something in the air. And

Norgren recalled seeing a rotor cloud not a half hour before, about halfway between Colorado Springs and Pueblo to the south.

Georgia Matteson's ears pricked up at the mention of her own aircraft when Rayfield told the United jetliner, "Traffic eleven o'clock five miles northwest bound is a Cessna seven thousand one hundred straight in for runway 30." The Boeing 737, United Flight 585, replied, "OK, we'll look for him, ah, how many miles are we from him?" Rayfield replied 5 miles, and was asked the same question about 25 seconds later. This time he replied that the Cessna was no longer a factor, since the small plane was now well behind the much faster jetliner. At 9:42 A.M., Rayfield cleared another aircraft for takeoff, then gave Matteson's Cessna final clearance to land. At 9:43 and 33 seconds, Matteson confirmed her instructions with the tower. Ahead, and below and to the right, she noted Flight 585 lined up for its final approach. Then, almost before she knew it, she saw the jetliner "heading straight down." From his seat beside her, Norgren also spotted the horrifying sight. Matteson was that day emphasizing the importance of constantly scanning the forward view to avoid collisions, and what Norgren saw at this moment was the underbelly of a jetliner as the plane made what looked like a vertical dive toward the ground. Almost simultaneously, Rayfield's excited voice erupted over the radio. He yelled "crash" seven times in succession, once saying "crack" in error, as he raised the fire crews by radio.

Over the radio Matteson asked the tower, "You know what's causing the black smoke over there?" Of course, she already knew what had happened, and was stunned, and felt helpless and useless. She even offered to remain at altitude to relay information to the tower, to help, to do *something*. But Rayfield tersely replied, "Just continue to runway 30."

After discovering to his relief that the Fountain Valley School had been spared, Detective Crouch sped to Widefield Park to see if he could help with the rescue effort. From a block away he saw a lot of smoke, and some flames. And at the park itself he saw six or eight official vehicles—fire trucks, ambulances, and police cruisers—already pulled up on the grass at crazy angles, with approaching sirens signaling lots more were on the way. And of course he saw a crowd starting to gather—maybe he could best help just by keeping them at a safe distance.

But there was no sign of the elegant silver, white, and dark-blue airplane he had seen disappear beyond the ridge. It took a little time for him to realize why: the jetliner was almost totally buried in a crater in the ground, a crater of its own making. The fuselage had telescoped to one-tenth of its original length. Apart from fragments of wing lying on either side of the smoking hole, and the short protruding stub of the tail, there was precious little evidence of the 60-ton, 100-foot long, $14-million aircraft that had been United Flight 585.

The pieces of the jetliner that weren't buried were twisted and shattered and torn and blackened by fire—most past the point of recognition. Detective Crouch quickly realized that no survivors were going to emerge from the hellish wreck. He recruited a friend, Gregory Roberts, and a uniformed motorcycle patrolman, Russ Weiss, to keep onlookers away from the site. By now the emergency crews had what Crouch didn't consider a very large fire under control, but he still wanted the crowd kept far away to fend off souvenir hunters who might descend on the wreckage before official crash investigators had a chance to secure the scene. After the park was cordoned off with yellow plastic police tape, Crouch and Roberts joined the fire teams in the gruesome task of marking and covering the human body parts that had been ejected on impact. These, too, were mostly unrecognizable.

Looking at the sixty tons of a Boeing 737 lumbering along a taxiway, its wings flexing alarmingly under the weight of fully loaded fuel tanks, it is hard to imagine how the huge machine will ever get off the ground. Yet most of us (phobics aside) don't give flying a second thought. Delays and missed connections, irritatingly high prices or bargain fares, awful food or the fabulous view—these we take almost for granted. But how and why do we take for granted sitting in a massive but thin-skinned aluminum tube at 30,000 feet?

Evolution started the production of beautiful living flying machines many millions of years ago, but humans, who will always need artificial aid to fly, only mastered the technique within the past century. Apart from balloonists, who couldn't steer their craft, early attempts at manned flight went up a blind alley by imitating birds. Success came only after pioneers abandoned flapping in favor of rigid wings.

The first heavier-than-air flight carrying a pilot in the United States took place in California in 1883, when a Santa Clara University professor, John Montgomery, flew 600 feet in a glider he had designed himself. It was also the world's first controllable aircraft. In 1896, a group led by French-born Octave Chanute experimented with glider designs in a camp they established among the sand dunes of Lake Michigan's southern shore, but their longest flights lasted less than a minute and were usually far short of 1,000 feet.

Nevertheless, these early aviators were proving that airplane flight was a coming thing years before the first motorized flight to carry a human, the Wright Brothers' famous flight in the *Flyer* at Kitty Hawk in 1903. Incidentally, the Wright Brothers cannot lay claim to the first powered flight. That credit goes to astronomer Samuel Langley who, in 1896, made a model aircraft powered by a miniature steam engine that flew for more than a half mile before running out of fuel. His failure to perfect a steam engine light enough to power an aircraft large enough to carry a man is possibly all that prevented him from grabbing the Wright Brothers' place in history.

The key to the success of the early gliders, and the Wright Brothers' *Flyer*, was the special curved shape of the wings. For that they had a German glider pioneer, Otto Lilienthal, to thank. Lilienthal put a higher curve on the upper surface of his wings, which achieved greater lift than anyone imagined possible. (Lift is the upward force produced by air flowing over a wing.) Lilienthal glided further and higher than anyone else until he was killed in an 1896 crash. Before his work, the first experimental glider wings were barely curved, and in aerodynamic terms were little better than a kite.

To understand how a wing produces a lifting force, imagine that you are looking at an airplane wing from the side. You would notice that the bottom of the wing is relatively flat, but the top is curved so that it bulges upward slightly. When the airplane is sitting on the ground, the air pressure is the same on the top and bottom surfaces of the wing. As the airplane begins to move, air flows over and under the wing. The air flowing under the wing flows straight past the wing, and the air pressure on the bottom surface of the wing is largely unchanged from when the airplane was stationary.

However, the top surface of the wing presents a different story. The flowing air is deflected upward by the curved surface and has to travel farther to get past the wing than the air flowing under the wing. Because the air has to travel farther, its speed increases. Over 200 years ago, the Swiss mathematician Daniel Bernoulli theorized that the pressure in a fluid decreases as its speed increases. What this means for a wing is that the air pressure on the top surface is less than the air pressure on the bottom surface, and the wing is pushed upward. By moving through the air, a wing has created an upward force that we call lift. The faster an airplane moves, the greater the lift. As an airplane accelerates down a runway, the lift increases until it is great enough for the airplane to take off. For both birds and airplanes, lift is necessary for flight.

Early aviators discovered, often fatally, that taking off posed few terrors in comparison with controlling the plane in flight. Apart from a simple skid or runner, the *Flyer* had no undercarriage and it moved so slowly that one of the Wright brothers ran alongside and kept up with it for some of its historic flight. Aside from a rudder, there were none of today's controls that precisely direct an aircraft's movement through the air. For example, modern aircraft have ailerons, small hinged panels at the back, or trailing edge, of each wing that can move up or down to keep the plane level or to dip one wing or the other to cause a turn. In place of ailerons, the Wrights simply added a mechanism to the *Flyer* whereby the pilot could twist the wing, a rather crude effort to achieve the same result.

If getting into the air was the beginning, staying there would tax the intellects, and claim the lives, of pioneer aviators for many years to come. Until each successive breakthrough was achieved and mastered, pilots would die as their aircraft stalled, spun, broke up, caught fire, suffered engine failure, ran out of fuel, and misbehaved in ways they could not have foreseen. During World War I, the number of airmen killed by accidents, malfunctions, and bad design and engineering easily outnumbered the fatalities from aerial combat.

By the end of World War I, aircraft design was creating shapes that we find largely recognizable today. By the 1930s, wood and stretched fabric biplanes were being replaced by more reliable metal monoplanes.

The Boeing 247, a 12-passenger aircraft launched in 1933, might not have been a commercial success, but it pushed aircraft design down a route that led, inexorably, to the Boeing 737. The 247 was one of the first US-built twin-engine airliners with a stressed metal skin, which gave added strength to a lightweight aluminum framework. "They'll never build them any bigger," proudly remarked Boeing's chief engineer as the first 247 was wheeled out. But it was rapidly eclipsed by competitor Douglas Aircraft Company's DC-2. This faster, larger aircraft evolved into the famous DC-3, which was the workhorse of the US armed forces during World War II. The DC-3 became a hugely successful commercial airliner after the war—just as the commercial aviation passenger business began to grow by leaps and bounds.

Boeing made a dramatic comeback in 1957 with the Boeing 707 four-engine jetliner. Although originally designed as a military transport, it outsold Douglas's DC-8, which closely resembled it. These jetliners traveled faster and higher than ever before because they borrowed heavily from their high-performance military cousins. Wings were now angled back like jet fighters' in what is known as a swept-wing configuration, as opposed to being fixed at right angles to the fuselage. Boeing scored magnificently again in 1967 with the 747, the first jumbo long-range jetliner. Its medium-range 727, launched in 1960, was already becoming a classic aircraft. Boeing now had an aircraft for almost every long- and medium-haul niche in the jetliner market. However, when Douglas announced the DC-9, a narrow-body twin-engine jetliner seating more than 100 passengers, Boeing realized it had nothing to offer airlines seeking a jet replacement for smaller propeller-driven short-haul aircraft.

Using parts of the fuselage and other components from the 727, the 737 was created to quickly fill the gap in Boeing's product line. But it was far from being a quick fix. For the fourth time in a row, Boeing had created a jet-age winner. The new short- to medium-haul jetliner had a stubby-looking fuselage and a seemingly oversized tail. While it lacked the sleek good looks of the DC-9 or the 707 (it was nicknamed the ugly duckling), it more than compensated for its ungainly appearance in economy, reliability, and safety.

The 737 started off with just one airplane design, but over the years since the first aircraft was delivered to German airline Lufthansa, it has grown to include a family of airplanes of different sizes. The first group of 30 aircraft, designated the 737-100, seated up to 100 passengers, but with the success of the new plane it was almost immediately supplanted by the longer 737-200, which could accommodate up to 130 people in a longer fuselage. The even longer 737-300 first flew in 1984, and the family has continued to grow with so-called New Generation and Next Generation aircraft, some smaller, some even larger, taking the series up to an -800 version, with even further variants on the Boeing drawing board.

Like its British and French competitors, British Aircraft Corporation's 1-11 and Aerospatiale's Caravelle, the DC-9 had its two engines mounted in pods on either side of the rear fuselage. The 737 designers opted for an engine slung in a pod beneath each wing, about halfway along its length. Early passenger jet engines were extremely noisy. While distancing engines from the fuselage improved passenger comfort somewhat, it created other problems. First, because they were slung beneath the wings, the engines increased the size of the side profile of the aircraft and created more resistance to winds from the side. This lead to trickier handing on takeoffs and landings in crosswinds. And there was another problem with this type of engine mounting: should one engine fail, the other, being so far out from the fuselage, created more torque, or turning moment, than a fuselage-mounted engine. The result was a tendency for the aircraft to fly in circles. Boeing solved both problems with a larger rudder, the largest of any aircraft its size. Other aircraft, especially those with fuselage-mounted engines, had smaller rudders mounted on their tail fins, and these were sometimes split horizontally into two sections. An airline pilot normally has little use for a rudder except in emergencies or when he experiences a strong crosswind near the ground when landing or taking off. In most aircraft, one section of the split rudder was sufficient for everyday needs. The other section of the rudder was a back-up in case the first rudder failed. But the 737 designers were worried about the turning moment of one engine if the other engine failed, and decided that half of a split rudder was not enough. They wanted to make a single-panel rudder large enough to compensate. A

737 pilot flying on one engine could use this larger rudder to counteract the turning moment of flying with a single engine.

Pilots may use quick movements of the rudder to get their aircraft under control in difficult situations, such as landing or taking off in strong crosswinds. By and large, they don't normally use them to steer. Ailerons, hinged panels at the back, or trailing edge, of each wing, are used instead. They have the effect of causing one wing to drop and the other to rise, and the aircraft rotates around the dropped wing. Ailerons produce a subtle movement of the aircraft called a roll, a gentle lean to one side or the other that is less unsettling for the passengers. Large rudder movements, on the other hand, can be quite dramatic because they can cause the tail to swing violently from side to side, a process known as yawing. They can also cause a severe roll. Keeping the rudder turned can lead to a crash because the roll becomes so severe that the nose of the aircraft points toward the ground.

The rudder on the 737 has one more function, which goes largely unnoticed by most pilots. At higher altitudes, where the air is thinner, swept-wing aircraft like the 737 sometimes develop an unfortunate condition called Dutch roll, which is a tendency to roll and yaw from side to side in flight. (The phrase was coined by early aviators because the oscillations it produced were said to be like the swinging motion of a Dutchman skating down a frozen canal.) Dutch roll can make passengers very uncomfortable. It can be easily corrected by minor movements of the rudder. But rather than have pilots constantly work the rudder pedals at their feet, a type of autopilot called a yaw damper operates entirely automatically to correct the phenomenon. When the yaw damper detects a yawing movement, it sends an electrical signal to the rudder controls. The result is a tiny, imperceptible twitch of the rudder that swings the tail back in line with the aircraft's intended course and cancels out the Dutch roll. In the vast majority of cases, yaw dampers work so subtly that passengers are unaware of Dutch roll or that gentle rudder movements are taking place, sometimes hundreds of times an hour. But yaw dampers had problems.

In 1969, after a series of complaints from airlines about faulty yaw dampers, Boeing announced that it had discovered a manufacturing fault in an electrical connection in the yaw damper circuit that may have resulted in unnecessary rudder movements. In June of that year it issued a service bulletin that described how to test the connection. But the problem still did not go away.

One airline that was particularly affected was Frontier, one of Boeing's earliest US customers for the 737. In 1970, Frontier agreed to test a new yaw damper that Boeing claimed would eliminate the sort of problem the airline was experiencing. However, the following November, Frontier reported that one of its 737s developed a "severe roll rate" just 50 feet above the runway when landing at Denver. Such a roll so close to the ground was dangerous because a wingtip could hit the runway before touchdown. A month later, in an urgent telex to all its 737 pilots, Frontier announced that yaw dampers were to be switched off when landing and departing. This edict was to remain in force until a fleet-wide test of yaw dampers could be completed.

The issue was more fully spelled out in a "must read" letter sent the following day to all pilots by Frontier's manager of jetliner training, who added that the damper was also to be switched off any time abnormal yaw developed. That letter, which said malfunctioning yaw dampers were causing "serious control problems," also revealed that another incident had occurred 11 days earlier on December 10, 1971, aboard a Frontier 737 at 16,000 feet.

Boeing was consulted, and recommended taking extra precautions to prevent moisture from getting into the area known as the E & E bay, the electrical and electronic bay. This bay is the compartment where many vital components, including the yaw damper, are located. Boeing experts said there was growing evidence that moisture could seep through the floor from the galley and from rainwater collecting where the front steps fold into the door frame. That meant moisture could collect on yaw damper electrical connections, causing unpredictable and intermittent short circuits that send random commands to the rudder. Frontier pilots were now told not to use any autopilot equipment if they suspected moisture had invaded the E & E bay. They also learned that 737 aircraft with faulty yaw dampers sometimes gave a noticeable single yaw before the damper failed completely—a handy way of knowing when to switch it off.

In January 1972, Frontier announced an accelerated yaw damper overhaul program with the aim of replacing all faulty components in just four months. But the yaw damper problem was not so easily solved. On August 16, 1973, a Frontier crew was flying their 737 with the autopilot off at cruising altitude when a violent yaw occurred. As is normal, the yaw damper was switched on. The incident was reported when the aircraft landed at Bozeman, Montana. The yaw damper was labeled "inoperative," but the aircraft was still allowed to return to service without maintenance. This led to an investigation by the Federal Aviation Administration, which found that returning the faulty aircraft to service was a breach of its rules, and Frontier was fined $2,000.

The hard luck continued for the Frontier 737 fleet. On August 22, 1973, Captain Max Schow and First Officer Donald Strauss, flying the same plane involved in the August 16 incident, experienced a violent yaw. They were on autopilot at cruising speed and about to descend from 26,000 feet on their flight between Bismarck, North Dakota, and Rapid City, Iowa. As Strauss described it, "The aircraft abruptly turned to the left, and rolled to the left and slightly down, as if a hard left rudder had been applied." Captain Schow switched off the autopilot and completed the flight to Rapid City. The flight attendants at the rear of the aircraft were flung about by the incident but were not seriously injured. Captain Schow continued the next leg of the flight to Denver, where a maintenance supervisor finally grounded the aircraft, citing that a short circuit had caused the problem.

And there was another incident. On January 22, 1975, Frontier Captain W. R. Hurt reported that his 737 had experienced a sudden left movement "as though someone had pushed full left rudder." Captain Hurt disengaged the autopilot and regained control, but one flight attendant was advised to get a hospital checkup after being thrown hard against the forward galley door. A week later, Frontier's vice president for operations decided enough was enough, and declared that despite the costs, he was proposing to accelerate a yaw damper modification program and complete it before the year end: "Although accelerating the program will cost an additional $60,000, we believe it is necessary for the continued comfort and safety of our passengers," he wrote in a memorandum.

The urgency of the order and the extra expenditure, as it turned out, were appropriate. A month later, on February 24, 1975, a Frontier crew experienced a sudden sharp yaw shortly after taking off. Maintenance engineers tested the yaw damper. They found no faults, but the problem recurred when the aircraft was taken for a test flight. With the problem persisting, but no flaw apparent, the entire yaw damper had to be replaced. What was troubling to Frontier was not just the time and expense of replacing components, but the growing sense that failure seemed more and more widespread in their fleet, and that it was unexplained. No one, it seemed, knew quite what was causing the problem.

Despite dedicating serious resources to the issue, Boeing never completely solved the yaw damper problem. Over the next few years, various fixes were tried but the problem never fully went away. Some aircraft were completely immune, but rogue rudder movements continued to erupt intermittently on a handful. Eventually, Boeing reduced the yaw damper's rudder movement to just 3° in either direction so that the danger would be minimized. In any event, many 737 pilots seemed to develop a sixth sense, often switching off the yaw damper well before it could cause trouble. They never for a moment imagined that such a small defect could affect the record of what had become one of the world's safest jetliners.

Until subsequent events refocused attention on the 737 yaw damper, the issue largely faded from view. When safety investigators were later probing the history of 737 rudder problems prior to 1990, they found that very few airlines had bothered to make an issue of yaw damper malfunctions by formally logging them as incidents requiring investigation by federal authorities such as the Federal Aviation Administration or the National Transportation Safety Board. The yaw damper issue had become something that airlines and their pilots had decided they could live with. Anyway, there didn't appear to be much point in complaining about a problem that even the world's largest aircraft manufacturer couldn't solve.

On February 25, 1991, what appeared to be a yaw damper incident took United Airlines First Officer Dave Allen by surprise. Allen was flying a Boeing 737 with tail number 9299 under the command of Captain Sandy Bebee between the two Illinois cities of Peoria and Moline. Flying at an airspeed of 280 knots, he had just leveled the aircraft after climbing to 10,000 feet on the short 20-minute hop when the aircraft yawed as if the right rudder had been applied.

Bebee recalled that it felt as if someone had slowly stepped on the rudder pedals for one or two seconds. He immediately switched off the yaw damper and the flight continued without further incident to Moline. As usual, it was flight attendants at the rear of the aircraft who had most felt the effects of the sudden yaw. "We had to cling to each other for stability ... the event happened so suddenly," reported Senior Fight Attendant Pat Marsalis, who was seated beside a colleague when it happened.

Two days later, Marsalis was again working on the same aircraft, number 9299, when there was a repeat incident. This time the flight was between Springfield, Illinois, and Denver. First Officer Ron Evans was at the controls, climbing in mild turbulence at 280 knots when he experienced two or three "very abrupt" small rudder movements that he at first did not think significant. Five minutes later came a steady, smooth rudder movement. Evans corrected the yaw with the rudder pedals and switched off the yaw damper. He and Captain Jim Donovan discussed what to do, and they decided not to climb above 30,000 feet as planned, but descended to 22,400 feet, where the denser air suited flying without a yaw damper.

After the first incident aboard aircraft number 9299, United mechanics replaced the yaw damper coupler, the component that detects and corrects the almost imperceptible deviations in heading that herald the onset of Dutch roll. After the second incident, the rudder transfer valve, a hydraulic component in the tail that accepts electrical commands from the gyro, was replaced. Four days later, on March 3, 1991, at 9:43 and 43 seconds in the morning, United Airlines Flight 585 plunged into Widefield Park in Colorado Springs, killing all 25 passengers and crew aboard. Its tail number was 9299.

Chapter 2 TIN KICKERS

Investigators from the National Transportation Safety Board are sometimes called "tin kickers," and watching them work on a plane crash, it's not hard to see why. They walk around the site, sometimes gingerly lifting pieces of metal with their toecaps, then letting them back down again. Sometimes they will stoop or squat and turn over an item of wreckage, and if they look puzzled or bewildered, it's because they usually are. Here, amid the still-smoldering, barely recognizable remains of what once was a precision-built piece of flying machinery, is probably the last place they are going to find the immediate cause of a crash.

There are, in general, two widespread misconceptions about the NTSB. The first is that it investigates only aviation accidents. In fact, the mission of the organization, which is a 600-employee federal agency independent of all other arms of government, is to monitor the safety practices and investigate the breakdowns of safety in all modes of transport—including the trucking industry, freight and passenger railroads, bus lines, airlines and private aviation, shipping, boating, and even oil and gas pipelines.

The other widely held misconception is that the investigators we see on newscasts, searching for the familiar "black boxes" or reconstructing the wreckage in a nearby hangar, pretty much represent the sum total of what the agency does in the aftermath of a crash. But this could hardly be further from the truth. After the crash of United Flight 585, to be sure, it was only hours before the first NTSB crews were on the scene at Widefield Park. By nightfall that Sunday, an advance party of eight had already flown in from Washington, D.C., and within a few days a total of 18 NTSB people were assigned to the investigation in Colorado Springs. But in fact, this very visible group of investigators was augmented within days by dozens of other experts, drawn from Boeing, United Airlines, the Federal Aviation Administration, the Air Line Pilots Association (ALPA), the International Association of Machinists and Aerospace Workers, and various component manufacturers, including Pratt & Whitney, which made the doomed aircraft's engines. Each organization becomes what is called in NTSB parlance a "party," and each has a standing invitation to contribute personnel to any investigation in which it has an interest. Party members outnumber NTSB staffers in an investigation, often by as much as ten to one, and bring a breadth and depth of expertise that the safety board, from within its own ranks, could never match. But the parties *do* have interests—be they financial, public relations, even emotional—in the outcome of the investigation. They are not "disinterested" observers, and they bring more than just their expertise to the scene.

Critics of the party system often comment how, as sometimes happens, a party member tries to influence the direction of an investigation or attempts to draw attention away from something he would rather not see probed any further. But NTSB staffers always defend the system, which, they say, helps them get quickly to the heart of how an aircraft works. "If you can come up with a better way, please tell me," is the stock response from investigators when asked if they are comfortable with the collaborative arrangement with "outsiders."

Within the NTSB, staff investigators come from many backgrounds. Of course, aeronautical and mechanical engineers predominate in any air disaster investigation, but there are also ex-pilots, retired air traffic controllers, psychologists, and even former flight attendants. Indeed, even to say that the NTSB is conducting *a* crash investigation is something of a misstatement, since there are usually several parallel probes taking place, each with its own focus, each with a separate leader, each drawing on a wide and deep reservoir of expertise—and each, as it turns out, with its own interests, its own outlook, its own preconceptions, and often its own conflicts. Coordinating all of these investigations is the NTSB investigator-in-charge, a senior manager and one of a select group of experienced professionals who undertake the overall leadership role in major investigations on a rotating basis.

At Colorado Springs, Al Dickinson, a Vietnam veteran and one of the elite group of senior investigators, was picked to fill the role of investigator-in-charge. Older hands in the NTSB keep a suitcase permanently packed in the office or in the trunk of their car for just such an eventuality, sometimes even with a winter and a summer layer included, just to cover all the possible sites they may have to rush to. Wives and husbands have long ceased to be surprised to kiss their spouses goodbye in the morning and then that evening to get a phone call from hundreds of miles away, or even overseas, saying they won't be home for dinner.

Dickinson was in the downtown Washington area that Sunday when his pager called him to report to the NTSB. He contacted colleagues who were also on standby duty that weekend and arranged to meet them at Washington's National Airport, where they boarded a Gulfstream executive jet belonging to the FAA. Within a well-rehearsed matter of hours, he and what the NTSB calls a "Go Team" were on their way to Colorado Springs. Joining the team on its flight was John Lauber, a former NASA psychologist who in 1985 had been appointed to the board—the panel of five presidential appointees who oversee the NTSB—by Ronald Reagan. The five members of the board all have standby duty, but usually only travel to the scene of a major accident. Unfortunately, what had happened in Colorado Springs met that qualification. Lauber was to be the public face of the NTSB for the next week, holding the nightly press conferences and making some administrative decisions, and leaving Dickinson and his investigators more time to get on with the job.

The first official meeting of the whole group was at the Radisson Hotel in Colorado Springs, just 13 hours after the accident. Dickinson was accompanied by several colleagues from the NTSB who were tapped to head up groups examining different aspects of the accident, and together they quickly went through the advance list of other specialists from the parties to the investigation who were already en route to Colorado Springs. Most of them would show up at the incident center being established by Dickinson within 24 hours of the crash.

Later, the first major administrative hassle behind him, Dickinson traveled through the darkness out to the crash site with John Lauber and several of the investigators, some of whom were ready to start work immediately. He wanted to see what would occupy the best part of the next year and a half of his life with his own eyes. The cordoned-off area of the park was illuminated by arc lamps to help in the search for human remains and wreckage. This was his first large aircraft accident investigation, and he wanted to familiarize himself with the scene. Before being called to spearhead the 585 crash investigation, he had worked on nothing larger than a commuter aircraft accident, and he recalled being surprised at how little wreckage there appeared to be for him to evaluate. "It all came to rest in the one big hole," was the best way he could sum it up.

It was eerie, the strangely still and almost empty patch of brightly lit park surrounded by houses and apartment buildings. And it was all the more so when Dickinson returned to the park some 18 hours later. Most of Monday had been taken up with assigning various party experts, who were flying in from across the nation, to their appropriate groups on the investigation team, and the next time he got to see the wreckage, that evening, it had snowed. Everything was covered as with a white shroud.

Predictably, media attention quickly focused on retrieving the aircraft's two black boxes, the flight data recorder and the cockpit voice recorder. These talismans of air catastrophes—which are in fact orange, not black—may be critical, of course, but they represent just one small part of any investigation. What's more, however telling the black box information may be, as often as not the data and recordings they contain do not shed significant light on the causes of the accident. And nowhere was this more true than in the case of United Flight 585.

All the major networks were covering the crash, and Lauber faced a battery of cameras, lights, and microphones at his first news conference late on the Sunday night after his return from the crash site. Inevitably, he was asked about the search for the black boxes, and he was able to announce the discovery, just minutes earlier, of the cockpit voice recorder, which as he spoke was already en route to Washington for analysis. He was still asked, "What caused the crash?" after that revelation, as though the answer were scrawled on the outside of the battered recorder. It would be Thursday before the first words trickled back from the CVR specialists in Washington. Dealing with the media was always a delicate balancing act for Lauber. He knew it was safety board policy to be as open as possible with the media, but he still had to be careful about what he told them. He didn't want to get ahead of the investigation, especially if it would mean he had to issue a retraction or a clarification afterward. Trying to reel back statements could cast doubt on NTSB's credibility.

He had already foolishly told reporters that parts of the aircraft were buried up to 30 feet deep in the ground, a detail he would be correcting within days. In fact, all he could reliably say was what investigators would be doing. Speculating on the cause of the accident was a matter for others, and there was plenty of that going on already. When reporters asked him if the weather, and particularly the reported rotors, might have had a bearing on the accident, he replied that reconstructing the weather was one of the things the NTSB did automatically.

One might reasonably think that black boxes aboard airplanes are a fairly recent development, but that's not the case. The history of flight data recorders (FDRs) is as old as powered manned flight itself. Even the Wright Brothers fitted a rudimentary device to their *Flyer* to document its propeller rotation, the distance traveled, and the time spent in the air. And some two decades later, Charles Lindbergh took a more sophisticated version aboard the *Spirit of St. Louis*. Encased in a plywood box, it inked its data, including altitude, on a paper roll. But it was by no means crashproof.

Since July 1, 1958, federal regulations have required most US airliners to carry an FDR that continuously records four flight parameters—airspeed, altitude, heading, and vertical acceleration—as a function of time. It was a requirement that should have been operational a decade earlier, but until the late 1950s the aircraft industry had failed to produce an instrument capable of withstanding the forces of a crash (100 Gs, or 100 times the force of gravity, was specified) or the heat of a 1,100°C flame for 30 minutes.

These early (that is, 1950s-era) recorders used individual recording heads for each parameter and produced oscillograph tracings on a foil roll that rotated at the painfully slow rate of 6 inches per hour. This could result in all the data covering an accident being contained within one-tenth of an inch of foil, no joke to the hapless technician who had to decipher it through a microscope. And this, of course, assumes the investigator actually had the data in the first place. This was not always the case because the early recorders were most often fitted into the cockpit or a well in the main undercarriage, parts of an aircraft likely to be severely damaged in an accident. In the early 1960s, new rules moved the FDR aft, nearer the tail, which statistically suffers less damage during impact. They also required ten times greater impact tolerance, 1,000 Gs, and greater protection against penetration, crushing, and immersion in fuel, oil, and seawater.

These early foil recorders were not perfect, but they had some advantages. Consider the French alternative, which used light-sensitive paper that could not withstand much heat, and from which the recording completely disappeared if inadvertently subjected to light before the images could be properly developed. The British developed a recorder that used wire as the recording medium; it tangled and broke easily in a crash. Neither model proved very practical, but the British system did at least have one advantage: it used a kind of digital technology to encode the data, which meant that very little recording medium was needed to hold what was essentially a compressed message.

Digital technology was embraced by the Federal Aviation Administration in the early 1970s, a period that also marked a changeover from foil or wire to magnetic tape, which was capable of recording digitally up to 25 hours of flight data. Cockpit voice recorders (CVRs), capable of recording 30 minutes of cockpit conversation and radio transmissions with air traffic controllers (as well as ambient sounds heard in the cockpit) made their debut at the same time. The CVRs added an extra dimension to what investigators could scrutinize about the critical few minutes before a crash.

Early digital FDRs converted analog, or variable, electrical signals from special sensors into a series of 1s and 0s, but new digital electronic cockpit instruments introduced in the early 1980s meant that the latest generation of FDRs could simply plug into cockpit systems to extract almost limitless data about an aircraft's performance. (The digital information already existed in the aircraft's various information systems, so no conversion was necessary.) This proved useful to some airlines, especially those in Europe, which made it a practice to download vast amounts of data from a secondary type of FDR called a Quick Access Recorder (QAR) and use it for maintenance troubleshooting, ensuring that their aircraft were being flown as safely and economically as possible.

Recorder technology took another major leap forward in the late 1980s with the introduction of solid-state digital flight recorders. Rather than having data written to a tape or disk, these systems stored the information in memory chips. The complete absence of moving parts in these latest machines meant crash investigators were less likely to have to first sort out a spaghetti-like mess of tangled tape before they could get access to their data. And with each generation of recorders, regulators added additional must-monitor parameters to a list that has now grown from the original 5 (including time) to cover 88 flight and cockpit-command events and conditions.

All of which was of little consolation to Tom Jacky, one of NTSB's recorder specialists, when he received Flight 585's battered FDR. Although recorder regulations had advanced pretty much in tune with the technology—as the machines got more sophisticated, what they were asked to record became more detailed and complex—there had been no requirement by the FAA that airlines fit newer machines into older airplanes. If you were flying an older airplane, chances were that it featured an older type of recorder which was "grandfathered" for the life of the plane. Although it was a digital device and used a high-capacity tape medium capable of recording several additional channels, Flight 585's recorder, a Fairchild Model F800, was set up to record just the four basic flight parameters plus time. In other words, it recorded only the five parameters that had been required of recorders for more than 30 years.

Flight 585's FDR had suffered greatly from the impact of the crash. Its external cover was extensively damaged and had to be cut away to expose the contents. The internal electronics were also badly damaged. When the internal, shockproof protective casing was opened, Jacky saw that the cover over the tape was broken and that the tape itself had come off its spool and was crumpled in a pile. Fortunately, he noted with some satisfaction, it was not torn or mangled, but he still faced the delicate and tedious job of carefully winding it by hand onto an empty spool before placing it in one of the NTSB's special recorders in Washington. The strategy would then be to play the original tape through until he could identify the accident sequence, then copy that portion of the recording onto a blank half-inch tape that would be used for all further processing.

The contents of Flight 585's FDR tape was absolute gibberish, like the contents of any FDR tape are gibberish unless you know what you are looking at. The tape contained rows and rows of numerical values from 0 to 4095 that had to be converted to feet, knots, and other aeronautical values. Jacky started by producing conversion formulas he derived from information supplied by United Airlines. These were then programmed into an NTSB engineering workstation, which quickly made order, if not sense, of the data. The tape, Jacky found, provided a good—but not perfect—record. Each parameter was measured just once every second with the exception of vertical acceleration, which was sampled 8 times per second, and altitude, which was sampled only once every 4 seconds. One could, of course, wish for a lot more raw data, thought Jacky, but at least a coherent story was beginning to emerge from the tape.

For most of its last minute of flight, Captain Harold Green's airplane had a heading of 302° magnetic—in other words, it was pointing slightly to the west of a northwest direction. Then, 7 seconds from impact, the airplane swung around until it was pointing more north and lining up with Runway 35 at the Colorado Springs Municipal Airport. It was at that point, Jacky determined, that something strange began to happen. The nose rapidly swung back to the south until it was pointing in almost the opposite direction, and then the nose veered slightly back toward the north again when the plane crashed.

Flight 585's airspeed (the speed of the aircraft through the air, not over the ground) remained reasonably steady between 145 and 175 knots for most of its last minute of flight. That is, until 6 seconds from impact, when the airspeed increased rapidly from 160 knots to its final recorded value of 213 knots. Jacky's figures also showed that during the last minute, Flight 585 had been coming down gradually from an altitude of 8,000 feet above sea level (the Colorado Springs airport is located on a giant plateau, 6,172 feet above sea level) until, 7 seconds before impact, the rate of descent increased rapidly.

Also, during most of that last minute, the G forces (vertical acceleration) experienced aboard the doomed airplane ranged between 0.7 and 1.3. The lower value was probably not dissimilar to the mild weightlessness experienced in turbulence when, as an aircraft suddenly loses altitude, whatever one is holding, like a beverage glass or a book, momentarily feels lighter and may even be involuntarily raised a few inches. (Until turbulence was more fully understood, this sort of experience was often described as "hitting an air pocket," as though the air there was somehow thinner.) The larger G force from the tape, 1.3, might be the rough equivalent of what passengers experience when an aircraft bottoms out of one of these dips. That is to say, the G forces data for most of the last minute showed some definite bumps, but not catastrophic changes. In the final 7 seconds, however, the G forces increased from 0.7 to 4.09, over 4 times the force of gravity.

Based on the readouts from 585's FDR, all Jacky could say with some certainty was that what the aircraft did in its final seconds was not so different from what was reported by some of the more reliable eyewitnesses back in

Colorado Springs. Somehow, as it made its sudden deviation from course, it had rotated onto its back, then spun and executed a vertical dive into the ground. But there was nothing in the limited amount of data he could extract from the recorder to say why a perfectly good, well-maintained airplane flown otherwise faultlessly by an experienced and highly commended crew should suddenly corkscrew out of the sky and crash, almost nose first. The *what* was clear, the *why* was not.

Dickinson's investigators had fallbacks. There were two other sources of recorded information: the cockpit voice recorder and, from the ground, the now-impounded tapes of radar signals showing Flight 585's last moments in the air. A radar site near Denver, the Parker installation, about 53 miles north of the crash site, was tracking Flight 585. Parker feeds its signals to controllers in several locations. This type of installation is called a Secondary Surveillance Radar (SSR); it monitors aircraft by picking up transmissions from an automatic beacon (or transponder) aboard each plane. This transmits the aircraft's identification signal (or "squawk"), its altitude, and its heading, all of which appear together in a cluster of numbers on the radar operator's screen. United 585's squawk code was 5154. Another NTSB team, the Recorded Radar Study Group, chaired by John Clark, was given its readout from the FAA's database of radar records. (Radar stations are operated by the FAA as part of its nationwide air traffic control network.)

Unfortunately for the investigators, there was disappointment in store. The radar antenna at Parker sweeps slowly through a complete circle and during each rotation updates the data on every aircraft within its range. The last transmission it picked up from Flight 585 was 10 seconds before the aircraft commenced its crash dive. Prior to that, the Parker SSR showed 585's gradual descent to Colorado Springs from 8,000 feet until it intercepted an electronic signal called the glideslope. This signal is generated by the Instrument Landing System (ILS) on the runway and shows up as a pink cross on a cockpit instrument called the artificial horizon. By keeping this cross centered and lined up between two reference symbols on the artificial horizon, a pilot can locate the runway and start to descend at a comfortable angle of 3°, even in poor visibility or at night. The last sweep of the radar antenna showed that 585 had descended below the glideslope, by no means a cause for serious worry, given the turbulent conditions that morning. In any event, the weather was exceptionally clear and visibility was more than 100 miles. Captain Green would have been unlikely to lose sight of the runway.

The other backup available to Dickinson's team was, of course, the cockpit voice recorder. Because it was such a short flight, the CVR, designed to record the final 30 minutes of a flight, had recorded almost all of the flight deck conversations during Flight 585's final journey from Denver. However, it too was not a perfect recording. Like the Flight Data Recorder, the device, manufactured by Sundstrand, had also sustained serious impact damage. The tape was creased in several places and, because it had been ejected into a creek, it also sustained water damage. The fidelity was poor. The recording commenced as the aircraft was taxiing out to the Denver runway, and it demonstrated the pilots' preoccupation with that morning's weather forecast. Despite clear skies and bright sunshine, they noted that the forecast threatened very blustery, turbulent conditions en route, and the possibility of northwest gale gusts at Colorado Springs.

"It's a nice-looking day," Captain Green commented. "Hard to believe the skies are so unfriendly." That provoked a discussion about the hazards of Rocky Mountain weather and what it could do to an airplane. Captain Green was flying the aircraft, and by convention this meant that First Officer Trish Eidson handled radio communications, navigation, and the checklist procedures that crews complete at various stages of a flight.

First-time visitors to a cockpit, perhaps reared on romantic stories of barnstorming fliers, are often surprised at the number of times airline pilots haul out a clipboard or a ring-bound manual and read out lists of instruments and conditions to be checked, and tasks to be performed. There may be a checklist before they push back from the gate, another while they taxi to the runway, another prior to takeoff, another after takeoff, and still another before they reach cruising altitude. Modern jetliners even offer pilots a foldaway desk on which to prop up all this procedural documentation.

On a longer flight in a Boeing 737, a cruising altitude of more than 30,000 feet would have been normal. But this morning's trip was a short hop, and Flight 585 went no higher than 11,000 feet above sea level. In one sense, its shortness made it an easy flight, but it also was one that took 585 through some of the worst of the turbulence in the region. Captain Green remarked that he'd never flown to Colorado Springs without getting sick. As they commenced their approach, First Officer Eidson obtained a weather update, and it confirmed the earlier forecast and relayed reports from other pilots of severe turbulence between 18,000 and 38,000 feet throughout Wyoming and Colorado. It also added that low-level wind-shear warnings were now in effect at Colorado Springs Municipal Airport.

Until its effects on aircraft were accurately analyzed by Professor Theodore Fujita of the University of Chicago in the 1960s and 1970s, wind-shear accidents were poorly understood and were often put down as pilot error. They mostly affected aircraft landing or taking off when airspeed was low and the aircraft was vulnerable to a stall. Aircraft land and take off best when they are pointing into the wind, which increases the flow of air over their wings and improves lift. But if the wind suddenly changes direction and intensity, that lift can be lost, especially if the change is also accompanied by a downdraft that may cause the aircraft to lose even more altitude. A skillful pilot can escape if he quickly recognizes the situation for what it is and applies more power. But if wind shear is forecast, careful pilots will add power in advance, increasing their landing or takeoff speeds.

Aboard 585, Eidson and her captain wisely decided to add 20 knots to their landing speed, bringing it up to 135 knots. This would give them a comfortable margin of airflow over the wings, and consequently a comfortable margin of lift, in case they should encounter wind shear. About 5 minutes later Eidson is heard asking the Colorado Springs control tower if any other aircraft reported gains or losses of airspeed, sure indicators of wind shear. The tower responds with information relayed earlier by another 737, which reported it lost 15 knots at 500 feet, gained 15 knots at 400 feet, and gained 20 knots at 150 feet.

"Sounds adventurous, thank you," Eidson replied. To her captain she said, "OK, I recommend we hold whatever 20 knots max is, and then if we get all stable, I'll watch that airspeed like it was my mother's last minute, and I'll report to you." Some 4 minutes and 25 seconds before impact, Eidson told Green he was lined up with the end of the runway, and Green replied, "OK, we're not going to be in a rush because we want to stabilize it out here." Eidson agreed. Then Green called for a flaps extension, and the sound of the flap lever on the center console between the two pilots is heard. Modern aircraft wings are designed to be at their most efficient at relatively high speeds, but a landing at such speeds would be very dangerous, if not impossible. Yet, if the aircraft slows down too much, the wings soon lose lift, leading to a stall and crash. Flaps, which are panels that are hydraulically pushed out to extend the rear surface area of the wing and thus increase lift, reduce the speed at which the aircraft stalls and allow for a slower landing.

At 3 minutes and 45 seconds before impact, Captain Green announced the start of a wide turn that would take the aircraft back in the opposite direction, for an upwind landing on Runway 35. He also called for the undercarriage, or landing gear, to be lowered. The clunks and whirrs of this taking place could be heard on the CVR tape, despite its poor quality. Then the tower called to warn them there was a Cessna in their neighborhood, but heading for a different runway. About 40 seconds later, Green called for a further wing-flap extension, to the Flaps 15 position, and the sound of this happening is on the CVR. Then Eidson rapidly went through the final-descent checklist, which was completed successfully, followed by Captain Green asking the tower for an update on the Cessna. He was told it was no longer a factor, as it was now behind him.

After another 20 seconds, Green called for a further flaps extension, to Flaps 25, and this was followed by some post-landing taxiing instructions from the tower and the sound of the engines increasing power, followed by Green saying, "Starting on down." Then there is the sound generated by the Colorado Springs Middle Marker, an electronic beacon to help aircraft line up precisely on the runway, and Eidson's voice saying, "The marker's identified, it's really weak." Up until and including this moment, the voice recording has been more or less routine, and for the next 40 seconds nothing even remotely out of the ordinary is said by the pilots as they focus on the final approach. Eidson suddenly exclaims, "A 10-knot change there!" to which Green replies, "Yeah, I know, an awful lot of power to hold that airspeed." Eidson then comments that their runway is 11,000 feet long, and about 20 seconds later comments on "another 10 knot gain." Then, 38 seconds before the end of the recording, Green calls for "Flaps 30," and this is followed by the sound of the flap lever being operated. The command "Flaps 30" allows a further reduction in airspeed, normally to about 110 knots in the case of a 737 carrying Flight 585's estimated load, but to which had been added at least 20 additional knots of airspeed to cope with wind shear. So all told, the plane at this point had an airspeed of approximately 130 knots. About 33 seconds before the end of the tape, Eidson emits a loud "Wow," but there is no additional explanation or comment from either pilot. Engine power is then reduced again, and there is no conversation for the next 20 seconds until Eidson, as is routine, calls out "We're at a thousand feet." Then she says, "Oh God, flip..." as the captain calls for Flaps 15. The word flip was not spoken clearly, and was partly interrupted by the captain's flaps command. It is now 7 seconds to the end of the recording, and Eidson responds "Fifteen ... oh!" The captain also yells "Oh!" and both pilots then swear forcibly. There is the sound of the flap lever being operated several times, the captain shouting "No," Eidson swearing again, the captain swearing again, Eidson twice yelling, "Oh my God," and the captain yelling, "Oh No," and swearing just before the sound of impact. And in those last few seconds is heard the heart-rending scream of Trish Eidson just before she becomes the first female American pilot to die in the cockpit of a scheduled commercial airliner.

Chapter 3 LOOKING FOR ANSWERS

Pilot error is a common cause of crashes but by no means the only factor, or even the prevalent factor. If one searches the history of aviation, or even just uses one's imagination and common sense, the causes, even in a greatly simplified accounting, can add up to a dizzying list of possibilities. Airplanes also crash because of bombs, fuel tank explosions, birds being sucked into jet engines, lightning strikes, ice forming on the wings, mid-air collisions with other aircraft, electrical fires, cargo fires, murder, suicide, engine explosions, engine failures, collisions with mountains, hills, and airport buildings, structural failure, power failure, hydraulic failure, severe weather conditions, just plain running out of fuel, or a microscopic flaw in a vital component that has gone undetected for years. Human factors play a large role: mechanics forget to fix things, pilots get lost or program their computers incorrectly, or air traffic controllers give wrong directions. And in the worst conceivable cases, as we all saw on September 11, 2001, human intruders can overpower a crew and seize the controls, turning airliners into deadly weapons of terror. Sometimes, it seems, the pilots do everything by the book, but it turns out the book was wrong or incomplete or didn't anticipate every eventuality. Investigators need to consider any and every possible factor, and of course that doesn't rule out causes they might not even have imagined.

Normally NTSB crash investigators, after making an initial visual inspection of a crash scene, will withdraw to allow the rescue services and the local coroner's staff to document and remove the remains of the dead. But Flight 585 had crashed in such a way as to require heavy construction equipment—backhoes, cranes, and grapplers—to dismantle and extract the wreckage piece by piece, effectively following the aircraft into the hole it had dug for itself. If Al Dickinson and his investigators were not to miss possible clues that might otherwise be disturbed or overlooked in this process, they would have to work side by side with those locating and removing human body parts. In no respects was this an easy task. Dickinson had flown Huey helicopters with the First Cavalry in Vietnam and was no stranger to seeing bloodshed and mangled remains. But the crash in Colorado Springs was another matter. In the course of his 15-year career at the NTSB, he had never seen this level of carnage. It wasn't just that his earlier crash investigations were of smaller commuter planes, but that the bodies had usually been removed by the time he arrived.

When an airliner like a Boeing 737-200 crashes nose-first at 213 knots, different things—invariably fatal—happen to its occupants. They are initially flung forward so their heads strike the rear of the seat in front, causing massive head and neck injuries. Their ankles and wrists similarly strike parts of the seat in front, causing fractures. As they shoot forward, their lap belts bite hard into their abdomen, causing serious injury to internal organs and possible pelvic fractures. In the minute fraction of a second it takes for this to occur, most have sustained injuries massive enough to kill them, or at least render them mercifully unconscious, because what happens next is just as unpleasant. Overhead lockers have burst open and now cannon their contents at the backs of the unprotected heads of passengers; a laptop computer, a bottle of liquor, even a gift for a child becomes a lethal projectile. In another millisecond any occupied seats, designed to cope only with a force of 9 G, have started to collapse under the immense strain, and their mountings are tearing loose from the floor, which is itself buckling as tremendous shock waves ripple through the plane. Similarly, the aluminum skin and interior walls of the aircraft are shearing away from their supporting skeletal framework of angled metal alloy, and some of these materials are splaying into the cabin. The newly exposed jagged metal edges are added to the tangle of collapsed seat frames and cracked-open floor panels to create a giant mincer, into which is fed the bodies of the passengers. Few bodies remain intact, and the compression of the 60-ton aircraft to a fraction of its original length acts like a giant plunger, squeezing some human remains to the outside through breaks in the fuselage.

Flight 585's fuselage was once 100 feet long from nose to tail. Now it lay compressed entirely into a hole just 9.5 feet deep, 24 feet wide, and 39 feet long. In a large area around the central impact crater, small streamers of red and yellow tape fluttered in the breeze. The more numerous yellow ones marked aircraft parts; the red, body parts. On either side lay the broken remains of the wings, embedded in the ground, part in and part out of the crater. Their fuel tanks had burst open, and jet fuel poured into the crater, where it had fed a fire. Miraculously, apart from an eight-year-old girl standing in the doorway of a nearby apartment building, blown over and temporarily stunned by the shock wave of the impact, there were no injuries or fatalities on the ground. Sheriff Bernard Barry marveled at the local community's lucky escape—or if it wasn't luck, he thought, it was the skill and bravery of the pilot.

By late Monday, the day after the crash, both black boxes had been recovered. While their data were being analyzed in Washington, other groups had already started investigations in their own area of expertise. Three in particular, Powerplants, Systems, and Structures, started their work at the crash scene, working side by side with emergency workers, trying to ignore the horror all around them and focus, as best they could, on the job at hand.

The Powerplants Group, as its name suggests, looks at the wreckage of the crashed aircraft's engines and asks whether or not they played a role in the accident. The fundamental question they have to answer is whether the engines were working properly at the moment of impact. What remained of 585's two engines had to be carefully examined for any evidence of pre-impact engine malfunctions that could have provoked the fatal dive. Among the theories advanced at an early stage was the failure of one engine, coupled with the pilots' failure to act in time by correcting for the sudden asymmetric thrust. (If one engine suddenly fails, the remaining operating engine will send the aircraft into a spin unless the pilot intervenes.) Similarly, could one of the engines' thrust reversers, normally used to brake the plane when it lands, have deployed in flight? This would have resulted in a sudden reversal of the thrust of the engine, and an even more forceful tendency for the plane to be sent into an uncontrollable spin. (Less than three months after the Colorado Springs disaster, an Air Austria Boeing 767 crashed in Thailand, killing 223, when one of its thrust reversers deployed at cruising altitude).

Or could there have been, the powerplant investigators needed to ask, something calamitous that happened inside the engine? Could the engine have suffered an in-flight explosion, or disintegrated due to metal fatigue? A jet engine consists of a series of rapidly rotating fans and turbines sucking cold air into the engine, compressing it and forcing it further back to a narrow combustion chamber where it mixes with fuel and, under even greater pressure, is ignited. The force of the hot gas passes through more turbines, spinning them and providing power, and escapes, rocket-like, out the rear. Although much more reliable than older, internal combustion piston engines, jet engines can sometimes fail catastrophically. The engine rotates at such high speeds that the loss of even a single fan or turbine blade can lead to a chain reaction causing major damage inside the engine. Jet engines sometimes disintegrate explosively, flinging blades and other debris at high speed through the air. Should they collide with the aircraft, the results can be devastating. When an engine exploded on a British Airtours Boeing 737-200 about to take off at Manchester in 1985, the debris punctured a fuel tank and started a fire that killed 55 of the 137 people aboard. The rear-mounted engine disintegrated aboard a United Airlines DC-10 in July 1989, and debris sliced through the aircraft's hydraulic lines, almost totally disabling it and leading to the death of

112 of the 296 people aboard when it subsequently crash-landed at Sioux City. Ten years previously, another DC-10 stalled and crashed at Chicago, killing 271 people when an engine fell off the wing, also severing hydraulic lines. Could Flight 585 have collided with a flock of large birds, causing catastrophic damage to one or both engines? Even a single starling can dislodge fan and turbine blades and wreck the engine. A birdwatcher had seen geese pass through the Widefield Park area some 20 minutes earlier, but nothing around the time of the crash.

In Widefield Park, the engines were buried in the ground beneath the wings to which they had been attached. The left engine was almost 7 feet deep in the soil, such was the force of the impact, and the right was found three-quarters buried and violently disintegrated. Some parts of its combustion chamber were found 25 feet away, and the thrust reverser, a small hydraulic valve attached to levers and rods, was 140 feet away. A large fan-like turbine disc from the right engine was found 553 feet to the northeast, with swirling marks in a 25-foot path in the grass suggesting it had landed still spinning rapidly.

One of the first tasks of the powerplant investigator is to measure the angles at which the crashed plane's engines are embedded in the ground. This can provide a good idea of the aircraft's angle of impact. If the embedding angles of the right and left engines differ, it might suggest that one wing struck the ground first, slewing the other around to a different angle, or that the aircraft hit nose first but was yawing or rolling about its axis. United 585's left engine was at a 75° angle to the ground. What remained of the right engine was closer to 50°, opening up both possibilities. The Powerplant investigators also carefully examine the interior of the engine for evidence of damage indicating an in-flight fire or explosion. The turbine blades, for example, need to be examined. There are several different-sized turbines resembling multi-bladed fans at various locations in a jet engine. Their role is first to compress air, and then the fuel-air mixture, prior to ignition in a combustion chamber. It is possible to determine if there was a pre-impact engine failure or an explosion that caused blade damage in the turbines, or if the blades were rotating normally at impact and suffered damage only afterward. All the powerplant evidence pointed to nothing more than post-impact damage. The engines, it seems, were running when Flight 585 ploughed into the ground at Widefield Park because all the blades were damaged from contact with soil that had been sucked into almost every crevice.

The question then became, *how* were they running? It was not possible, from United 585's relatively primitive flight data recorder, to determine whether the engines were at full power, or idling, or somewhere in between. But information was available, frozen in time, in the aircraft's wrecked cockpit, if you knew where to look. Crash investigators found all the engine indicator dials from the wrecked cockpit instrument panel, and they were taken to a United Airlines workshop for examination.

Most of the cockpit instruments were well-mangled by impact forces. Cases were smashed and indicator needles were broken off. However, investigators were sometimes able, by carefully looking at the internal mechanisms of a shattered instrument, to find the stub to which the needle had been attached and arrive at a close approximation of its reading at the moment of impact. It was similar to the way that a policeman, in the era when all clocks had moving hands, could determine the time of death of an accident victim by looking at when the impact had stopped his watch. Even the face of a destroyed cockpit dial could provide clues. Often on impact, a dial's glass cover would be pulverized, and its indicator needle broken off and lost. But by carefully examining the face of the dial, which heretofore had been protected and pristine beneath its glass cover, investigators would sometimes be able to detect a small blemish or indentation made by the indicator needle at the moment of impact. (Interestingly, a more modern plane with purely digital instrumentation would not yield this kind of evidence, although it presumably would be equipped with a much more sophisticated flight data recorder.) As it turned out, Flight 585, with its older, more traditional instrumentation, did yield some readings. United Airlines technicians found several needle impact marks on the painted faces of the dials that reported engine status, and sure enough, the readings were within normal operating ranges at the moment of impact. So the Powerplant Group, after all, had little to offer in terms of clues, let alone a single compelling factor that could explain 585's plunge from the sky. The engines, all their evidence told them, were not the source of the problem.

The Systems Group, led by Joel Ryan, sought to determine if any of the aircraft's internal and external moving parts were at fault. These include things like the cockpit instruments, the machinery that operates the complex system of flaps, ailerons, and other movable surfaces on the wings, the elevators on the tailplane (the small rear wings that extend from either side of the back end of the fuselage), and the rudder. Light bulbs from warning panels were sent back to the NTSB's metallurgy lab in Washington for examination. The filaments had broken due to the impact, but investigators might discover something important from learning how they broke. A clean break suggested the filament was cool, and not lit to indicate a warning or a danger condition at the time of impact. A stretched filament, on the other hand, like a piece of gum pulled from the mouth until it snaps, would suggest a warning light was hot from being illuminated when the plane crashed. The exercise uncovered nothing worthwhile.

At the accident scene, Ryan's group concentrated on finding major hydraulic and flight control system parts for later examination. They found all of them, but the crash and fire had destroyed most hydraulic lines and rods connecting flight control systems to their actuators.

Was there a problem with the aircraft's overall structure? Did something fail due to metal fatigue, or midair collision, or the force of a catastrophic meteorological phenomenon? These are among the questions to which the Structures Group, led by Ronald Price, was seeking answers. One of Price's first tasks was to order an aerial search of the aircraft's flight path to see if any critical parts might have fallen off and precipitated the accident. Nothing was found. Neither was there any evidence of fire damage to vegetation or structures south of the impact crater, suggesting the fire had started after impact, not before.

All the groups took aircraft parts as they were removed from the crater and laid them in their relative positions on the grass beside the larger pieces of wreckage. Then, once they had been labeled and photographed, the whole lot of parts was dispatched to a hanger the NTSB had rented at the Colorado Springs airport. There, Price and his Structures team laid everything out again, looking for evidence of abnormalities other than those caused by impact damage. They found nothing.

Powerplants, Systems, and Structures all had their work cut out for them in Washington, at the airport in Colorado Springs, and at the site. But a fourth group—Operations—which had no need to go near the wreckage, possibly had the widest brief of any NTSB working team working on the crash. Chaired by Timothy Borson, a senior investigator in Washington, Operations was charged with establishing the background to the aircraft's last flight. Who were the crew, what experience did they have in flying the aircraft, were there any shortcomings in how the airline functioned with regard to training and briefing them? The Operations Group would attempt to reconstruct the aircraft's previous few flights to see if there were any flaws or anomalies that might offer clues to the accident. The group also assembled, in the minutest detail, a history of the final flight, including as much as could be ascertained from witnesses about its last moments. In the case of Flight 585, they even discovered the macabre fact that a coffin carrying human remains, a young man who had been killed in a traffic accident, was part of its doomed cargo. They also reviewed the flight history of the airplane and collated extensive information on techniques used to fly it. Unlike some other groups that dealt largely with post-crash wreckage, Operations adopted a more historical perspective. Was there anything in the employment history of the pilots that suggested that they might be prone to making mistakes? Were there any recurring glitches with the 737s in United Airline's fleet that made them tricky to fly? Had Boeing or the FAA issued any bulletins that affected the 737's operation and, if so, were the pilots made aware of them? Did the airline ensure that its pilots were checked regularly according to FAA regulations? Was the aircraft carrying any dangerous cargo?

Malcolm Brenner, an NTSB psychologist who was rapidly carving out a new specialty in aircraft crash investigation, aided Borson. He chaired an Operations subgroup called Human Performance. Brenner's investigation probed deeply into the personal and medical histories of the two pilots in a bid to unearth anything that might have contributed, however remotely, to the accident. For example, had either of the pilots any undisclosed medical conditions that might have triggered a heart attack or a stroke while at the controls?

Meanwhile, other sub-groups within Operations were methodically going about their work, trying to piece together the full story of Flight 585. As with others investigating operations, these remaining groups did not deal with wreckage, or even visit the site. Air traffic control expert Allen Lebo, a former air traffic controller himself, set out to interview James Rayfield, the controller on duty at the time of the accident, as well as Rayfield's colleagues, and to obtain the results of drug and alcohol analysis on the blood samples the FAA routinely requests from its employees, including air traffic controllers if they had any connection with an accident aircraft. Lebo also planned to review the tape recordings of the tower radio traffic, which had been sealed following the crash. Along with other researchers, but especially those in Operations, he would also have access to tapes of the aerial tracks of Flight 585 as detected by several radar stations along its flight path. John Young headed a probe into a virtual mountain of the maintenance records of the crashed plane, a study that by midweek had revealed details of the two earlier rogue rudder actions that, according to United's records, had been cleared up by the installation of a replacement rudder component five days previously.

Some groups, quite literally, were attempting to conjure evidence out of thin air. Meteorologist Greg Salottolo was painstakingly trying to assemble a picture of the weather at the precise moment that United 585 fell from the sky; reproducing the turbulent atmospheric chaos of that day was proving no easy task.

John Clark's Performance Group had the vital and, as it transpired, extremely difficult job of re-creating and explaining United 585's path through the air in its last seconds of flight. Unlike a train or auto wreck, an aircraft leaves no trace of its passing in the sky. In aviation there are no skid marks or ploughed-up tracks to follow. Sometimes, investigators use the analogy of police work at a murder crime scene to explain how they approach an accident and how their groups relate to each other. The Structures Group is comparable to the officers handling the scene of the crime; its "corpse" is the wreckage.

The various forensic experts also have their analogous groups. The Systems and Powerplants groups, for example, are among aviation's pathologists, probing the deceased to see what contributed to its demise. You can find fingerprint and other crime scene experts in the groups that look for evidence in the environment outside the wreckage—for example, the meteorologists and metallurgists. The Operations Group contains aviation's gumshoes, metaphorically pounding the pavement to assemble a complete profile of the victim, just as a cop interviews all known associates with the possibility of discovering a motive. The Performance Group recreates the victim's last known movements by pulling in data from aviation's security cameras, i. e., radar records and the flight data recorder. Aviation even has its wiretaps and phone records, in the form of the cockpit voice recorder and recordings of transmissions with air traffic controllers.

Over the coming months, as the different groups sent him their reports, or, if they were still investigating, updated him on progress, it became apparent to Al Dickinson that no clear cause of the accident had emerged. Borson's Operations Group and Brenner's Human Performance subgroup had assembled a picture of an exemplary captain and crew who did everything by the book. Lebo's Air Traffic Control Group reported that the controllers had been faultless in their handling of the airplane. Price's Structures Group found nothing in the wreckage to suggest a cause of the accident, and the trawl through United's maintenance records indicated that the airplane had been well cared for.

There's a saying among NTSB staffers that if they don't have a clue as to the cause of a crash within the first week, an investigation will take months, even years. The black boxes were yielding no smoking gun. The pilots had said nothing coherent about their dilemma to each other or to James Rayfield in the tower in the seconds before they died. One after another, promising paths for investigation seemed to come to a dead end. Al Dickinson was getting increasingly frustrated. After working for several years on light aircraft and commuter accidents, Flight 585, his first large aircraft accident as investigator-in-charge, was a pivotal point in his career as a crash investigator. But as group after group drew a blank, Dickinson was faced with the prospect that this could be one of the few unsolved air accidents in the history of the NTSB, and it would happen on his watch. This was above all else a human tragedy, not a technical or engineering one, but all Al Dickinson had was a large pile of badly smashed-up wreckage in a hanger. And very little to go on.

Flight 585 was bringing unique pressures to bear on an investigation, Dickinson began to believe. Or if not unique, then pressures he'd never before noticed. This accident involved one of the US's most popular airplanes. The usual steady trickle of inquiries from the traveling public was to be expected, for sure, but more intriguingly to Dickinson was the larger-than-expected number of inquiries from pilots who flew the 737. They were, to put it bluntly, spooked by the notion of an airplane simply falling out of the sky—without any obvious explanation, without any specially announced and explicit reassurance that the aircraft was safe to fly.

After leaving the army, Dickinson had, for a short time, flown as a commercial pilot for a small airline. He understood the pilots' anxieties. But he had no answers to give. The investigation would have to take its plodding course, every lead chased down and investigated, then either eliminated or added to a small and shrinking list of possibilities. Most of them were eliminated.

Over the next few days, Dickinson called several all-hands meetings where each group was assembled with all its members and asked to give a presentation to the other groups on its work to date. Such meetings were often the first time that group members who were not NTSB staffers got to learn what progress the other groups were, or were not, making. Like any brainstorming session, the intention was that a presentation might trigger an idea in the mind of a person in a different group. He had seen the process work before, but this time it was just an exercise in further frustration. Inevitably, apart from suggestions that the weather on the day might hold a clue to the crash, the best most people could come up with was a suggestion that the Structures and Systems groups perform further trawls through the wreckage in the hope of turning up more clues. Dickinson agreed with a heavy heart and not much hope that the exercise would produce results.

Any and all leads were followed up, no matter how remote they seemed, no matter how tasteless or gruesome or even lurid they appeared to be. There was, for example, gossip that the pilots, Harold Green and Trish Eidson, had been having an affair and had fallen out, and in a fit of jealous rage, Green had taken the fire axe from the cockpit and killed her, then pushed the plane into a suicide dive. Another version of that story featured a different jealous lover who burst into the cockpit and murdered both of them. Malcolm Brenner got the job of checking those stories out. There was not a scintilla of truth in either one.

During the week after the accident, Dickinson requested that local radio stations and newspapers broadcast appeals for more witnesses to come forward. And people living under Flight 585's flight path were asked to be alert for any hitherto undiscovered pieces of wreckage that might explain why this airplane mysteriously tumbled from the sky. Dickinson got a parcel from one Colorado Springs woman who said she found its contents near the park where the plane crashed. It was rushed to the NTSB's laboratory where it was identified as a part from a domestic refrigerator.

As the weeks passed, two tantalizing leads worthy of further investigation gradually emerged from the growing labyrinth of material in Flight 585's docket, the centralized summary file the NTSB opens on every major crash. The first came from the Systems Group. As part of their probe, members of Joel Ryan's team had dismantled and minutely examined dozens of hydraulic components recovered from the wreckage. Although some of these parts were badly damaged, there was little to suggest they had been operating other than normally. But there was one exception. Examination of a part used to move the rudder revealed there had been some very serious rubbing in an unlubricated area where metal-to-metal contact was unexpected. In this case, the rubbing was so severe that galling, the transfer of metal from one part to another due to a kind of welding process, had occurred. Could this part, the standby rudder actuator, have seized up at a critical moment in the flight, jamming the rudder in a dangerous position and causing the accident? Jean Bernstein, one of the NTSB's team of skilled metallurgists, was given the job of finding out.

Then there was the equally intriguing issue of the weather. The turbulence on March 3 had been unusually severe, and there were several reliable witness reports of unusually strong rotor activity. Could the crash of 585 be weather-related, could it have been caused by these whirling bundles of air? There were precedents, as Greg Salottolo was about to learn in the process of assembling a thick file of data on other accidents that had been linked to serious turbulence.

If the members of the investigation teams had been lukewarm about, or just confused and unconvinced by, any of the causes proposed so far, one expert party decided at an early stage that weather was the only horse worth backing in this race. That party was the Boeing Company.

Boeing had assisted in dozens of air crash investigations involving Boeing 737 aircraft in the US and elsewhere throughout the world. Nowhere, its delegates repeatedly told Dickinson and his colleagues, had they ever encountered anything like the crash of Flight 585. There must have been something going on in the sky above Colorado Springs the day of the crash. Freak weather, they suggested, remained a fertile avenue for further investigation.

On March 5, 1966, residents of Gotemba City, about 50 miles southeast of Tokyo, were surprised to see an aircraft trailing an unusual white vapor high in the clear blue sky above them. The vapor seemed very different from the regular parallel contrails normally seen streaming from a jet's engines—it was somehow cloudier, more mist-like. It could even be seen coming from the wingtips, away from the engines. Then pieces were seen to fall from the aircraft. One large cloud of vapor emerged and turned black. More puffs of vapor were seen. One man managed to snap a photograph that when enlarged showed the ghostly grainy image of what appeared to be a tailless, engineless airliner with a bent wing, streaming white vapor and plunging out of control toward Mount Fuji.

With the benefit of hindsight, it is easy to inject omens into the story of BOAC Flight 911. On taking off from Tokyo's Haneda International Airport, it had taxied past the still-smoldering wreckage of a Canadian Pacific DC-8 that had crashed on landing the previous evening, killing 64 passengers. Operated by British Overseas Airways Corporation, a state-owned ancestor of British Airways, Flight 911 was a four-engine Boeing 707 jetliner on the Tokyo to Hong Kong leg of an around the world trip. Just prior to takeoff, the aircraft's captain, Bernard Dobson, unexpectedly called the tower and requested a change from the flight plan he had only just filed. Presumably for the entertainment of his passengers, he wanted to take a detour past Mount Fuji, Japan's snow-capped sacred mountain, plainly visible from the airport in the clear, blustery air even though it was over 60 miles away.

Captain Dobson climbed to 17,000 feet, then started a gradual descent toward the mountain—and to his doom. The wreckage and the bodies of the aircraft's 124 occupants were found scattered over an area more than 1 mile wide and 10 miles long. Although they found traces of metal fatigue in a small area of the tail fin, Japanese investigators decided this had nothing to do with the otherwise inexplicable catastrophe. They were stumped. The airplane's flight data recorder had been completely destroyed in the fire that consumed what remained of the forward fuselage, and there was no cockpit voice recorder fitted to the aircraft. There had been no emergency communication from the crew; whatever caused the aircraft to break up in midair had been sudden, and very violent.

Investigators found paint from the distinctively colored tail fin on one of the tailplanes, suggesting that the fin had collapsed to the left, breaking off the left tailplane in the process. Then the rest of the tail tore away from the fuselage, and the engines broke off the wings, which themselves started to break up. And the destruction of the wings fractured the fuel tanks, the source of the white vapor seen by witnesses on the ground. The collapse of the tail fin alone could have precipitated the chain reaction that led to the mid-air breakup, but what could have snapped off something designed to withstand 140-mph gusts?

A clue was provided by the pilot of a Skyhawk fighter dispatched from a US naval base to search for wreckage soon after the alarm about Flight 911 was raised. As he flew near Mount Fuji, not far from the crash site, the Skyhawk pilot ran into turbulence so extreme he afterward said he feared his aircraft would break up. He banged his head several times off both sides of the narrow cockpit, his oxygen mask shook loose from his helmet, and the aircraft did not respond to controls. Back at his base, the readout from his flight recorder showed the aircraft had endured forces of nine times gravity (9 G) as well as minus 4 G. The fighter jet had been so violently buffeted that it was grounded for a complete safety inspection. And it was later learned that the Skyhawk's encounter had not been unique: other aircraft that had flown in the same area that day also reported severe turbulence.

Winds at the summit of Mount Fuji were blowing between 60 and 70 knots at the time, and weather satellite images showed clouds symptomatic of mountain weather effects downwind of the Susuka Mountains, 150 miles away, although none could be seen in the dry air over Fuji. Flight 911 had, it appeared, been struck by an incredibly strong and forceful gust of wind, so strong it exceeded the design limits of the aircraft's tail fin. The Japanese authorities investigating the crash concluded that it had flown into severe turbulence amplified by the mountain poking high into the 70-knot wind. Whatever had struck the fin was gusting well beyond 140 mph.

The fate of Flight 911 has become one of the classic examples of what freak mountain weather can do to an airplane. The case history was also one of the first obtained by Greg Salottolo as he commenced his review of aviation case histories for some clues toward a solution of the mystery of United Flight 585.

Colorado Springs lies at an elevation of more than 6,000 feet above sea level. On three sides, the country is flat and almost featureless. On the fourth, though, to the west, is the dramatic Front Range, a row of high mountain peaks that hide the foothills of the Rockies even further west. The range is dominated by the craggy bulk of Pikes Peak, which rises for more than 8,000 feet above the plain below it, and by Mount Rosa and Cheyenne Mountain, which flank it. One question Salottolo now asked was whether Pikes Peak and its fellows might have played a part in the downing of Flight 585. He delved into the archives to look at other accidents that had been blamed on weather coming off mountains, as well as one weather-related crash in which the nearest mountains were hundreds of miles away.

In 1964, a US Air Force B-52 bomber lost three-quarters of its tail fin and rudder when it was savaged by turbulence at 14,000 feet to the east of Spanish Peak, an outcrop of the Sangre de Christo mountain range located about 95 miles south of Colorado Springs. The crew luckily managed to bring it back safely and afterward recorded their experiences:

> ... The encounter was very sudden and lasted about 10 seconds. During the first part ... the airplane appeared to be stable in that it wasn't moving in roll nor particularly in yaw and there wasn't anything on the

> instruments that would indicate anything more than normal excursions. As the encounter progressed we received a very sharp edged blow which was followed by many more. As the first sharp edged encounter started bleeding off we developed an almost instantaneous rate of roll at fairly high rate. The roll was to the far left and the nose was swinging up and to the right at a rapid rate. During the second part of the encounter the airplane motions actually seemed to be negating my control inputs. I had the rudder to the firewall, the column in my lap, and full wheel and I wasn't having any luck righting the airplane.

The B-52 is one of the largest and most powerful military airplanes ever constructed in the US, yet it appeared in this account to have been tossed about the sky like a tiny, underpowered light aircraft. As its pilot wrote above, he applied full rudder and had pulled the control column fully back as well as fully twisting the wheel atop it. Yet the airplane failed to respond. At one stage it even appeared to be doing the *opposite* of what he was commanding. Yet, as Salottolo was to find as he continued to sift through the aviation disaster archives, there were even stranger weather-related incidents where nobody lived to tell the tale.

In Pedro Bay, Spotsy Lake, Alaska, in December 1968, all 39 died aboard a Wien Consolidated Airlines Fairchild F-27B when unexpected mountain turbulence ripped off the right wing and parts of the left wing, sending the airplane into a fatal spin.

Falls City, tucked away in the southeast corner of Nebraska, is hundreds of miles from any serious mountains. Its weather is often dominated by massive lines of thunderstorms rolling off the Great Plains, but Salottolo knew thunderstorms and mountains are both capable of generating dramatic downdrafts of cold air from aloft and have violent, rolling, rotor-like vortices of air in common. Falls City was added to his list because on August 6, 1966, all 42 people aboard perished when a Braniff Airways BAC 1-11 jetliner disintegrated in midair near that remote town. It had been trying to negotiate a line of thunderstorms, and analysis revealed that the pilot had erred in entering the wall of storms at too low an altitude, at a point where turbulence was probably at its worst. Violent vortices of air probably ripped off the tail fin and right tailplane, leading to the midair breakup of the remainder of the aircraft. Could similar atmospheric conditions in the lee of the Front Range have rolled Flight 585 into its fatal dive?

A gravel-voiced New Yorker with a graying mustache, Salottolo is a slight anomaly at the NTSB because, unlike many of his colleagues, he doesn't have a private pilots license. "I love aviation, but that kind of flying is too dangerous," he says dryly, with the air of a man who has scientifically weighed the odds and found them wanting. Fascinated by weather from an early age, Salottolo studied meteorology at New York University, graduating in 1968. After a spell as a weather officer in the air force, he joined a private company, preparing forecasts for executive jet operators.

When he was hired by the NTSB in 1978, it was on the eve of a golden age of understanding dangerous weather phenomena. Researchers had been probing the massive thunderstorms known as supercells and unlocking their mysteries. The most famous of them was Professor Tetsuya (Theodore) Fujita at the University of Chicago. Fujita first came on the aviation disaster scene in 1966, when he deduced that the Falls City crash had been caused by turbulence within the thunderstorm. He took his work a stage further in 1975, when, after poring over satellite photos of New York's Kennedy Airport, he identified the cloud structure responsible for the deadly vertical downburst of wind from a thunderstorm that had forced Eastern Airlines Flight 66 into the ground a half-mile short of the runway on June 24, 1975, killing 120 of the 132 passengers aboard the Boeing 727.

Fujita's theory of downbursts from thunderstorms was controversial at first. He postulated that Eastern 66 had been felled by an immense downdraft, yet local weather instruments recorded only light winds at the nearby airport. In addition, barely minutes before Eastern 66 crashed, several other aircraft flew through the same storm yet landed safely. Some meteorologists were skeptical that winds of a ferocity capable of downing an airliner could otherwise escape detection, but a succession of accidents under similar conditions soon convinced them. Fujita christened them microbursts, a word that conjures up the image of a localized, intermittent shaft of terrible vertical wind capable of delivering fatal hammer blows to aircraft on their final approaches—that is, aircraft that had already throttled back to the verge of stalling as they slowed down for a landing.

Fujita, whose interest in thunderstorms had started when he was a teenager in his native Japan, stumbled on the existence of downdrafts from thunderstorms in his studies of the Great Plains landscape, where they left dramatic fingerprints in the form of mysterious circles of flattened crops or, in some extreme cases, even felled stands of young trees. He noticed that the damage often radiated out from a central point, and he realized this could only be caused by a strong, localized shaft of wind coming straight down to the ground, more or less at a right angle. Hitting the ground, the wind, blowing at up to 145 mph, radiated outward in all directions like an inverted mushroom, flattening vegetation until the microburst lost energy. The phenomenon, confined perhaps to a few thousand square yards, lasted only two or three minutes and faded almost as quickly as it began.

To understand how violent and devastating a microburst can be during an approach, one only has to imagine the scene as it unfolds from the vantage point of the cockpit. The pilot of an aircraft about to land, moving toward an invisible microburst that lies directly between his plane and the runway, first encounters an unexpected headwind blowing toward him, away from the center of the downdraft, and reduces his engine power to maintain the correct airspeed. To the novice, this may not make sense—why *reduce* speed in such a situation? The answer is actually quite simple. A pilot regulates his speed according to the rate at which air flows past his airplane, not his speed over the ground, and the decision to reduce thrust to keep the airspeed within acceptable limits for landing is almost instinctive. A pilot coming in for a landing has to play a careful balancing act between reducing his speed for touchdown but maintaining sufficient airspeed to stay aloft.

And the aircraft's attitude, or pitch, as well as its speed plays an important role in such an approach. The pilot approaching a microburst, in response to the suddenly increased headwind, might also point his nose down a bit to counteract the newly increased tendency of the airplane to fly higher, above the glideslope, because increased airflow also produces increased lift on the wings. Then, as he emerges from the headwind and enters the vertical downdraft area, his horizontal airspeed decreases because the airflow over his wings diminishes as the headwind suddenly disappears, and he notices he is rapidly losing altitude as the downdraft forces him toward the ground. Now his gut instinct is to pull back on his control column to raise his nose. Then, if he hasn't already stalled or hit the ground, he exits the downdraft on the opposite side of that inverted mushroom and he experiences a tailwind. This is classic wind shear. His speed is too low, his nose is too high, and the angle of attack of his wings to the air is all wrong, so that even if he starts to quickly get his engine power back up, a stall is inevitable. He loses lift and the plane, with not nearly enough altitude to attempt a recovery, falls to the ground.

Salottolo worked on several crashes blamed on microbursts and was an early convert to Fujita's theories. Indeed, the ten years between 1975 and 1985 was, without exaggeration, the decade of wind shear and microbursts in the airline industry. Hundreds died in a succession of crashes commencing with Eastern 66 and culminating in the horrific crash of a Delta Tristar, Flight 191, which was forced down by a microburst and burned, killing most of those aboard, after crashing just short of the Dallas-Forth Worth airport in 1985. Landing accidents were the most common, but there were also takeoff accidents. Salottolo worked alongside Fujita in solving the crash of Flight 759, a Pan Am 727, on takeoff at New Orleans in July 1982, when all aboard and 8 people on the ground were killed after a microburst-induced wind shear prevented the pilots from gaining altitude, and their 727 slammed into houses.

During that period, and in the years following, the NTSB issued almost 100 recommendations, ranging from calls for the installation of better instruments to detect and measure wind shear to dissemination of vital information about microbursts to pilots and control towers. And, while there would be exceptions, it appeared that the subsequent partnership program between the industry and relevant government agencies had finally licked the wind shear and microburst problem. Pilots learned how to recognize when they were flying into deadly danger and how best to wriggle out of it. But most of all they learned to treat thunderstorms with a new respect.

"Up till then, people didn't recognize what they were. I guess they called them air pockets, turbulence, that kind of stuff. Eastern 66 at Kennedy might have been the beginning of recognizing what was going on," Salottolo recalled. "The whole wind shear thing gave us all the greatest satisfaction. There was a very large effort between industry and government to lessen those hazards, and we were very successful. It's a model of how we should tackle other weather-related problems like turbulence and icing."

The question remained, however: did mountain-effect weather produce the same kind of terrible downdrafts? Was the crash of Flight 585, Salottolo wondered, just one of many instances of mountain downdrafts that were now just beginning to be recognized? Suddenly, to his alarm, it seemed to him that responsibility for solving this crash was somehow devolving onto his shoulders. In meetings, representatives of Boeing, the world's largest aircraft manufacturer, were nodding knowledgeably in his direction as he spoke about the weather, and he realized that, apart from some still unfinished business concerning the crashed plane's tail, everybody else was running out of theories and looking his way.

Chapter 4 RECONSTRUCTING THE WIND

Twenty-four days after the accident, Greg Salottolo and John Clark, Chairman of the NTSB's Flight 585 Performance Group, convened a series of meetings of outside meteorological experts in Boulder. Salottolo and Clark, along with their NTSB teams, wanted to pick the experts' brains about violent mountain weather phenomena and how they might have affected Flight 585.

The outsiders present included Al Bedard from the National Oceanic and Atmospheric Agency (NOAA) and Terry Clark from the National Center for Atmospheric Research (NCAR). Both had dedicated a large part of their careers to the study of what meteorologists call orographic flows—mountain winds—and their downstream effects. Also present were graduate students from the University of Colorado and Professor John Marwitz from the Atmospheric Science Department of the University of Wyoming. Laramie, where the University of Wyoming has its campus, is in the shadow of a Rocky Mountain range, like Colorado Springs. Since 1965, the Wyoming Atmospheric Science Department had operated a succession of twin-engine aircraft equipped as flying laboratories to study wind, turbulence, clouds, and temperature, among dozens of other parameters. In 1994, their studies were to prove central to the identification of drizzle as a producer of dangerous wing icing in freezing conditions. The department also had a deep interest in the microburst phenomenon, which it had been studying since the early 1980s. Its expertise would later lead it as far afield as Hong Kong, where it advised on the testing of an innovative wind shear detection system at the new airport—built, alarmingly, in the lee of a mountain.

The Flight 585 meetings moved between NOAA's facilities, the University of Colorado, and NCAR—all in Boulder. All the meetings seemed to convey the same message to the NTSB visitors: short-lived, transient mountain events could badly affect an aircraft that happened to be in the wrong place at the wrong time.

Al Bedard liked to remind newcomers to the Rockies that Boulder experiences three times more hurricane-strength winds than Miami. It's the best place in the entire US to get a free roof. Designers and construction firms seek out guinea pigs—people whose homes have lost their shingles in a windstorm—and test experimental systems and new roofing materials. Mountain airports, Bedard also reminds visitors, have a 50 percent higher accident rate for light aircraft than low-lying areas.

Graduate students from the University of Colorado presented a model of the rolling eddies of air called rotors and how they might drive an aircraft into a roll. The presentation was such a success that the students were asked by the NTSB officials not to show it to the media, at least for the time being. "It was so realistic I think they were afraid people might jump to premature conclusions," Bedard recalled.

Bedard himself focused on an elongated Front Range cloud that appeared on satellite photographs about the time of the 585 crash. The air was so dry that day that clouds were unexpected, unless they marked some specific meteorological structure like a low-pressure trough. A line drawn through the long axis of the cloud eerily passed through Widefield Park. Bedard is still not sure exactly what the cloud signified in terms of the crash, but it clearly indicated some kind of discontinuity in the prevailing weather pattern.

The day after 585 went down, the University of Wyoming's flying meteorology laboratory, a twin-turboprop Beech King Air, was dispatched to fly a succession of approaches to the Colorado Springs airport in an attempt to detect anything in the air capable of bringing down a jetliner. The department had already been in the midst of a study of Rocky Mountain winter storms when word of the accident reached them, so overnight they decided to extend their survey to Colorado Springs.

What they found was intriguing. Despite a strong wind still blowing over the Front Range from the northwest, as on the day of the accident, the airport was in what is called a wind shadow. Below 5,000 feet there were lighter winds than further aloft, often blowing in the opposite direction. But the junction between the lighter and stronger winds was marked by turbulence and rolling bundles of revolving air called vortices. Professor Marwitz delivered the findings at one of the Boulder meetings, but he concluded that the data did not suggest anything seriously threatening to flight performance.

The assembled experts also discussed mountain waves, produced by a rapid airstream crashing over a steep ridge, and the impact they can have on conditions in the lee, or downwind side, of the ridge. Rotors can form in the lee of a mountain ridge whenever a wind of more than 25 knots blows at a right angle to it. For aviators, rotors can be a serious problem: sometimes you see them, sometimes you don't. If there's sufficient moisture in the air, telltale wispy clouds may betray their presence, but just as often they offer up no visual cue. Another indicator of mountain waves are ranks of stationary, lens-shaped (lenticular) clouds that appear when cold air, having descended rapidly in the lee of a mountain, is warmed and rises, cools again and descends, and is warmed again and rises once more. The process can be repeated several times, and each time that the air reaches a certain altitude, moisture condenses out to form the characteristic stationary lenticular clouds. Cap clouds sited over the summit are another indication of mountain waves. Cap clouds are caused by moist air borne rapidly aloft by wind pressure on the windward slope of a mountain condensing and forming a stationary cloud above it, irrespective of the wind speed. (The physics of the formation of lenticular and cap clouds is similar, but they differ greatly in appearance: cap clouds are fluffy, lenticular clouds are smooth.)

Terry Clark, an enthusiastic Canadian-born meteorologist from NCAR, spoke energetically about rotors and what they really signify. Clark is very dismissive of what he calls the simplified scenario, in which rotor formation calls up an image of a boulder poking out of a brook, generating eddies in its wake. In this scenario, the boulder represents a mountain, the eddies represent rotors, and the rotors tend to exist only in the lee of the mountain ridge. Clark paints a much more complicated picture. He talks about gravity waves—sheets of wind soaring vertically above the peaks, then collapsing like a breaker at the seashore, often from as high as 2 to 5 miles in the sky. Rotors, Clark believes, are also generated at this great altitude, in some cases well away from mountain ridges, and they sometimes plunge to earth with dramatic consequences. Clark had reports from Colorado of car windows being blown out in parking lots by sudden bursts of strong wind coming from the east—that is, from the direction away from the mountains and opposite the prevailing wind.

"Bursts of turbulence can be seen hitting the surface 10 to 15 kilometers out in the plains, like someone was aiming mortar shells from the mountains," he says. "It is these isolated pockets of turbulence that tear off the shingles from a house, or sometimes damage the whole house. They sound like express trains approaching, and during an intense windstorm you will hear these things going overhead every 10 or 15 minutes. When they actually hit your house, you will really notice the difference between the more or less steady 50- to 70-mile-an-hour wind and the sudden onset of a gust, which is more like 100 to 140 miles an hour."

Clark's mental picture is further complicated by strong, narrow, almost vertical downdrafts of cold air that rebound, almost bouncing, becoming strong updrafts called "jumps." And his view of things is complicated even more by low-altitude winds funneling through mountain passes, like the Palmer Divide, almost due west of Colorado Springs.

If there was any single dominant theme for Salottolo, Clark, and their NTSB teams to take away from the series of meetings, it was that there was no simple answer to be drawn from the weather on March 3, 1991. Because the meteorological conditions had been so freakish and because the theories they'd been presented didn't form a coherent picture, Salottolo and John Clark asked Terry Clark and his NCAR colleague Bill Hall to use the computer program they'd helped develop to build a three-dimensional model of the March 3 weather at Colorado Springs. Their goal was clear enough: they wanted a better idea of what Harold Green and Trish Eidson had experienced at 1,000 feet.

Reaching the goal, though, was another matter. Terry Clark fed in the data he had, but there simply wasn't enough to kick-start the program and make a model of the March 3 conditions. Bedard's trough, confirmed by satellite imagery, was complicating matters by rerouting the upper airflow across part of the area covered by the program, and the computer couldn't handle it. However, by chance, Clark had earlier generated a model of a January 1989 windstorm that closely resembled conditions on March 3, and he extrapolated from this model to suggest what might have happened to United 585. The model had taken between 20 and 30 hours to generate on a Cray supercomputer, and it was, to put it mildly, a conceptual stretch. But there it was: a moving 3-D picture. The model suggested that a line of 90-mph updrafts snaked up and down the Front Range that day, and surged in and out of the plains between Colorado Springs and Boulder. The geography near Colorado Springs, in particular the Palmer Divide, confused things even further because of winds whistling through and complicating the "jumps." But all the elements, taken together, provided a plausible moving picture of the weather that fateful day.

Clark's final report concluded that it was possible for jumps, capable of producing rapid upward motion, to travel across the line of the main north-south runway at Colorado Springs. He was unable to say if such a jump actually occurred, or if it could upset an aircraft. But his report did conclude that, judging by the FAA's syllabus for commercial pilot examinations, there was little understanding of this type of windstorm required of pilots.

Earlier, Salottolo had tackled the routine post-accident work of reconstructing the weather from local records and checking it against the various forecasts and weather warnings that had been issued that day. Had the weather been better or worse than forecast? Was there anything the weather forecasters could have said or done to save Flight 585? By and large, he discovered two things. One, the winds had actually been diminishing in strength as the time of the crash approached. Two, the weather information being transmitted by the tower gave significantly stronger wind strengths than were being recorded on automatic equipment. When the equipment was checked, it was found to be accurate, and Salottolo never received a satisfactory explanation of the discrepancy between the instrument readings and the forecasters' information that the controllers passed along to the pilots. All told, thought Salottolo, the discrepancy didn't amount to much. Weather forecasting is an imprecise science at best, and pilots are used to getting reports that don't always pan out.

Salottolo also went looking for any witnesses who recalled the weather in the fateful minutes leading up to the crash. Most of us are obsessed with weather forecasts, but we have very little in the way of detailed weather memories. We want to know, it seems, little more than what's sufficient to prompt us to grab an umbrella, throw on a warmer coat, or slap on some sunscreen. To most people, a day is either sunny or cloudy, wet or dry, warm or cold, windy or still. Anything approaching even a roughly accurate, much less a precise, recollection of the weather on a given day quickly fades. And of course, one's sense of the weather is highly subjective; what's "freezing" to one person may be just "brisk" to another. From long experience Salottolo had learned that, except for an aviator or a yachtsman or someone who routinely takes a deep interest in the weather, the general public makes for imprecise meteorological witnesses.

Then again, one's recollection can be sharpened if something particularly exceptional happens. From Denver, where Flight 585 originated, he had reports of trashcans sliding back and forth along a street. Then there was Harold Darnell's account of his pick-up almost being stalled on Grinnell Street by a surprisingly strong wind. The timing in Darnell's recollection was especially intriguing because just after he felt the wind gust, he'd seen the smoke coming from Widefield Park, little more than a few blocks away.

Over the years Salottolo had developed a kind of internal grading system for the reliability of eyewitness reports. Ordinary civilians, however well intentioned (and sometimes precisely because of their good intentions), were more or less at the bottom of the scale. One had to listen carefully to them. They could deliver pure evidentiary gold. But an investigator had to maintain a healthy measure of skepticism. Then there are those who for one reason or another have some special knowledge of the weather, and are therefore placed in a higher witness category. Several of these witnesses gave descriptions of weather phenomena on the day Flight 585 crashed. A glider pilot reporting heavy tiles ripped from his roof; another glider pilot watching branches snap and car hoods clatter around in a junkyard; a former air force sergeant reporting his house shaken by booming gusts—all of them, in Salottolo's mind, worth listening to very intently. And from witnesses aloft, he mentally cataloged the reports of a Cessna pilot's wild ride, and a US Army major flying a Beech Super King Air and encountering terrible wind shear, at one stage losing 100 feet of altitude and 20 knots of airspeed. All of these reports, from the area of the crash and about an hour beforehand, had to be considered with the utmost seriousness.

One might assume, reviewing all of these dire accounts, that if ever the weather on any day was going to be a potential suspect, March 3 was that day. The problem, and in this investigation there always seemed to be a problem, was that not all the data squared with a weather-as-culprit explanation. Several other, more precise measurements of the local conditions were available, and they painted a much quieter picture. The National Weather Service operated an office on the west side of the airport from which it collected observations, including the data from a gust recorder near the center of the airfield. Records from this device indicated that the strongest gust on the airfield around the time of the crash was 14 knots, barely a stiff breeze. Indeed, during the precise minute of the crash it recorded a speed of merely 7 knots, about 8 mph. In addition, the FAA operated a series of sensors at the airport to detect rapid changes in wind strength and direction that can give rise to wind shear. The wind shear alert system prompts air traffic controllers to issue alerts to pilots whenever wind shear conditions are present so they can increase power to cope with a headwind that suddenly becomes a tailwind. The records showed that no alert was necessary.

Colorado Springs also possesses a municipal air quality monitoring network that, among other data, collects details of wind speed and direction from a network of sites. The Pinello site, about 3 miles northwest of Widefield Park and the closest to the scene of the crash, has an automated device that averages wind speed and direction in 10-minute segments. The two segments ending at 9:20 and 9:30 A.M., 23 and 13 minutes prior to the crash, showed winds in excess of 20 mph. But at 9:40, that average had dropped to 10.4 mph, and during the 10-minute segment in which the crash occurred, it averaged just 6.4 mph. However, the Pinello site did experience a 47° wind shift, from 327° (north-northwest) to 280° (almost due west), between 9:30 and 9:50 A.M. before swinging back into the northwest.

Salottolo commissioned a fuller analysis of the available weather data from a meteorological consultant. This analysis showed the wind shift had been along a line that passed roughly east-west through Widefield Park. At 9:40, winds on one side of the line were from the northwest at about 10 mph; on the other side of the line, they were weaker, about 6 mph, and coming from the southeast. Some 10 minutes later they had weakened further. Many witnesses near the crash location at the time said the winds were very light, or even that the air was still.

He also took the wind data from both the wind shear alert system and the air quality network and fed it into an NTSB computer using a program specially developed by the Space Science and Engineering Center at the University of Wisconsin. The analysis showed that the winds were converging on an area just south of the accident site. It also indicated they had been rotating around a wobbly vertical axis, almost like an extremely weak and loose tornado. But again, Salottolo found nothing known to be capable of bringing down a modern jetliner.

Armed with these extensive, if inconclusive, new sets of Colorado Springs weather data, John Clark's Performance Group turned to another important tool of the accident investigator—the flight simulator. From the outside, a simulator resembles a video arcade ride minus the flashing lights. Inside, it is a faithful mock-up of an aircraft cockpit, complete with controls, dials, lights, and even warning horns. The windows look out onto a large wrap-around video screen on which is projected the sort of scenery a pilot might expect to see at various stages of a flight. Simulators are usually modeled on a particular type of aircraft, and with software enhancements even variants of aircraft.

In the normal course of aviation, simulators are used in a wide variety of applications, including training new pilots, retraining pilots moving into a new model of aircraft, and rehearsing new flying techniques. Airlines also use simulators to check the proficiency of their pilots, who periodically have to be reexamined. Controlled by extremely complex and adaptable computer programs that over the years have evolved into extraordinarily sophisticated virtual worlds, simulators now reproduce the flight of an aircraft so faithfully that pilots value experience gained in them almost as highly as actual flying hours. The entire unit rolls, banks, and turns just like a real plane in response to the pilot's commands or to simulated weather conditions, while the instruments and the scenery react just as they would on takeoff, at cruising altitudes, in a holding pattern, or during an approach or landing. Engine noises and the whirrs and clunks of flaps and landing gear are fed into the cockpit to enhance the realism.

Because the financial commitment of owning a simulator for each type of aircraft flown in the US would be prohibitive, the NTSB, under the party system, makes extensive use of simulators belonging to the airlines and plane manufacturers involved in an investigation. Much of the investigative simulator work for the 585 crash was done at Boeing.

In April 1991, just a month after the crash, Jim Kerrigan, Boeing's representative in the Performance Group, ran simulations of encounters with typical rotors on Boeing's multi-purpose simulator set up to resemble a Boeing 737-200. He was also asked to see how the simulator might respond to a rudder hardover, a maneuver in which the rudder is pushed, accidentally or deliberately, as far as it will go to one side. The occurrence of a hardover was emerging as one of the few plausible explanations for the movement of the aircraft in its final few seconds in the air. John Clark took Kerrigan's hardover results, presented in graph form, and compared them to a graph of Flight 585's flight data recorder output. They did not match. The rudder hardover simulations made the aircraft's nose swing more quickly than it was shown to do on 585's FDR. Moreover, Clark and his associates could not persuade the simulator to force a weather-induced roll similar to the roll that 585 experienced, despite trying numerous sizes and strengths of virtual rotors.

Clark was also puzzled by the lack of altitude errors in the data. The higher in the air you go, the lower the atmospheric pressure. A plane's altimeter, the instrument that records altitude, is essentially a barometer that converts barometric pressure into altitude, and it can be fooled. Vortices, be they rotors or tornadoes, usually have an area of low atmospheric pressure at their core. A sudden excursion into a rotor's low-pressure zone would have tricked the altimeter into recording the aircraft as being between 150 and 300 feet higher than it really was. The problem with the Flight 585 recorder data was that it showed no altimeter errors of this kind: there was no momentary upward surge in altitude, but only the steady, slow curve of a descent toward a landing, and then the sudden plunge of the crash.

Some five months after the crash, on August 1, 1991, the Operations Group, under the chairmanship of senior NTSB investigator Timothy Borson, assembled at the Boeing simulator dedicated to the investigation in the company's plant in Renton, Washington, just outside Seattle. It was their third visit to Renton since the accident. This time, the purpose of the exercise was to see if the aircraft could be affected by the various in-flight "upset" scenarios postulated to date, and still be controllable. The simulated upsets ranged from rotors of various sizes and strengths to strong crosswind gusts to the sort of severe rotor phenomenon suspected of having ripped most of the tail off the B-52 bomber in 1964.

Mechanical failures and human errors were also to be tested. Could the Boeing 737 survive failures of moveable surfaces on the wings or pilot errors in extending or retracting them when required? These moveable surfaces could include the ailerons, flaps, and spoilers, which are hinged wing panels that can tilt up to disrupt airflow and reduce speed, as well as function as emergency ailerons. Could pilots recover from a rudder jam? This had been growing in importance as a possible cause of the accident ever since the Systems Group's discovery, four months earlier, of the damaged part in the standby rudder actuator. Had the galling discovered by the Systems Group and analyzed by NTSB metallurgist Jean Bernstein somehow caused the rudder to stick, even though it had occurred in a standby, or duplicate, unit that would rarely see action? A rudder jammed at its maximum deflection could cause an aircraft to turn upside down and plow into the ground if the pilots took no corrective action. But could this have been the case? The failure of a standby rudder component leading to a rudder jam in the hardover position *and* the failure of the pilots to correct it? This seemed beyond the realm of the possible.

And in any case, all the mechanical scenarios, even a rudder malfunction precipitated by galling, proved to be non-events in the simulator because they could easily be corrected using alternative flight controls. The various wind scenarios, even the rerun of the B-52 incident, failed to cause any severe flight control problems to the pilots in the simulator. Depending on how the simulator was programmed, an aircraft was sometimes kicked out of a rotor almost as soon as it entered. About the only rotor scenario that presented control problems was a situation where the aircraft entered the rotor and passed through it, not perpendicular to its axis but at a shallow 20° angle. After experiencing a roll in one direction as one wingtip, for example, was passing through an updraft, the pilot would attempt to get his wings level. Then, moments later, as the other wingtip experienced the updraft, the pilot would have to reverse his actions in order to stay on the level. Moreover, when the plane passed through the rotor, the whole series of actions and reverse actions would be repeated, but now with downdrafts instead of updrafts.

So the news from the simulator was more of the same—inconclusive. Nonetheless, Al Dickinson was still worried enough about the galling on the standby rudder actuator that he pushed for action. The standby rudder actuator was the part that, in the event of a failure of the main hydraulic systems in the airplane, actually moved the rudder. Like most standby systems, it was rarely, if ever, called into use. Yet, because of the complex interplay of rudder parts in the tail, it always moved in tandem with the main system, even if it were applying no force. Therefore, if any of its metal parts were in contact with each other, galling was a possibility. Similar galling problems, Dickinson knew, had been encountered on three other airplanes in the mid-1980s and documented in a Boeing service bulletin that also reported two of those aircraft experienced erratic rudder behavior. At Dickinson's urging, on August 20, 1991, the NTSB chairman James Kolstad wrote to James Busey, his opposite number at the Federal Aviation Administration, recounting the history of the investigation to date and recommending that he issue an airworthiness directive advising airlines to regularly check the 737's rudder standby machinery for traces of galling and to replace the parts if necessary. There was, Kolstad wrote, no evidence that this had led to the Flight 585 crash, but there was concern that in some cases it might lead to an uncommanded rudder input and subsequent control difficulties. The FAA obviously did not regard Kolstad's letter as urgent because it did not reply until early October, when Busey said the agency would issue a Notice of Proposed Rule Making—a document much less urgent than an emergency airworthiness directive. The issue was to drag on for the next two years.

A week before Kolstad wrote his letter to the FAA, Wayne Gallimore, a California aircraft mechanic and chairman of his union's local flight safety committee, wrote to Don Pomarico, a fellow mechanic and union official, to tell him about a conversation he had some days earlier with mechanics at the United Airlines hydraulics workshop. Pomarico was the International Association of Machinists and Aerospace Workers delegate on the Structures Group. Gallimore had earlier represented the union on the Systems Group, and as an IAMAW representative had been present at a March 1991 NTSB session on Flight 585 at the United Airlines' Maintenance Operations Center. He participated in the inspection of many vital parts removed from the aircraft, including a badly burned and damaged main rudder power control unit, or PCU. This precision hydraulic component mounted in the 737 tail translates movements of the pedals at the pilots' feet into rudder movements.

Gallimore's mechanics had reminded him of the condition of the PCU excavated from the wreckage of Flight 585 and how one of its moving parts was jammed against another part, within which it should have moved freely. The mechanics told him there were some surface nicks on the part similar to those on another PCU taken from an aircraft that had suffered a sudden, unexpected rudder movement while climbing out of Chicago the year before. Gallimore's letter to Pomarico, which Pomarico turned over to the Structures Group, had two main questions: Could a jam of those parts cause a rudder to swing uncontrollably to one side, creating a hardover? And could that have caused the crash of Flight 585?

"I believe that it is imperative that the NTSB be notified and a complete investigation be done immediately," the letter concluded. Nearly a year would pass before investigators realized that the letter was remarkably prophetic.

Chapter 5 WE HAVE A PROBLEM

On the afternoon of Wednesday, July 29, 1992, nearly a year and a half after the crash of Flight 585, Ron Schleede received a phone call that made him very angry. The caller told Schleede, who was chief of the Major Investigations branch of the NTSB, that a Boeing 737 rudder problem was being independently investigated by United Airlines—that is, the airline was conducting an investigation without reporting it to, or sharing any information with, the NTSB. Besides United, this "private" investigation involved Parker Hannifin, the company that produced many of the hydraulic components fitted to the 737. Parker Hannifin had just confirmed to United that under certain conditions a hydraulic component in the 737's rudder mechanism could unexpectedly force the rudder into maximum deflection, a so-called hardover. Schleede was furious that no one had thought this information was important enough to share with Al Dickinson, who headed the 585 investigation and reported to Schleede. Although there was no direct evidence that a full rudder deflection drove Flight 585 into its final inverted S-shaped curve into Widefield Park, everyone involved in the investigation knew, or should have known, that the "hardover hypothesis" was perhaps the simplest and clearest explanation for what had happened.

Questions about the 737 rudder were on the minds of Schleede, Dickinson, and just about everyone else involved in the investigation. Just two weeks before Schleede received the call about the "private" investigation, Mack Moore, a United Airlines 737 captain, was taxiing out prior to takeoff at Chicago's O'Hare Airport when, as part of a routine preflight procedure, he checked his rudder operation. He pressed the right pedal, and it felt normal. When he pressed the left pedal, though, it jammed about a quarter of the way down—he could push it no further. When he released pressure, the pedal returned to its normal position. In several additional checks during which he pressed both pedals harder than usual, he encountered higher resistance on the left pedal. Moore announced to the tower and his passengers that he was scrubbing the flight and returned the aircraft to the gate. United mechanics took the main rudder hydraulic component, the power control unit (PCU), from

Moore's aircraft and began testing it. These tests led to the private investigation that had infuriated Schleede.

Schleede was particularly irritated by the fact that Moore, just months prior to the incident on the runway at O'Hare, had been an Air Line Pilots Association representative on the NTSB's Cockpit Voice Recorder Study Group for Flight 585. He, of all people, should have known the importance of keeping the NTSB informed about pertinent developments. Instead, the details of the Mack Moore incident, as it came to be called, were kept from the NTSB for two weeks.

Previous NTSB probes into the rudder as a possible culprit in the 585 crash had focused on the standby rudder actuator, a 14-inch-long part that had been recovered in a mangled condition from 585's wreckage. The shaft emerging from the hydraulic unit, which pushes the rudder, had jammed against a bearing nut, leaving the telltale marks of galling. Because the standby actuator was a backup, and as such was rarely used in normal service, it was hard to work out how it could have caused an unwanted rudder hardover—if, in fact, this is what had happened. Tests had shown that the power generated by the aircraft's hydraulic system was sufficient to break out of the jam, although in the process friction generated so much heat that metal melted at the bearing and left deposits on the shaft. One theory was the remote possibility that when it was jammed, the standby actuator shaft, which should have moved freely whenever the main rudder hydraulics were in operation, somehow locked up part of the complex, interconnected rudder-control mechanism and created a rogue reaction in the rest of the rudder-control system. Earlier tests with the yaw damper had shown that when the standby actuator was artificially jammed to replicate the possible condition of 585's damaged actuator, a 3° yaw damper input was amplified to produce a 6° rudder deflection. This was anomalous, certainly, but still less than a quarter of full deflection, and therefore still not sufficient to render an aircraft uncontrollable.

Now, months after the rudder-control system tests, Schleede was discovering that the misbehaving PCU, or power control unit, from Mack Moore's aircraft demonstrated even more alarming tendencies. The PCU is not just a complicated device, but one that is in operation at virtually all times. Except in the rare instances where a pilot detects a fault in his two main hydraulic systems and switches them off, the main PCU acts as a sort of hydraulic switch, taking inputs from the rudder pedals (infrequently) and from the yaw damper (almost continuously at altitude) and translating them into left and right rudder movements. The PCU has to work nimbly and subtly, producing varying degrees and rates of rudder deflection. When the shaft is extended by the unit, it pushes the rudder to the left; when it is retracted into the unit, it pulls the rudder to the right. With an action analogous to power steering on an automobile, it can take a pilot input of 70 pounds pressure on a rudder pedal and convert it to 5,900 pounds of force on the arm that moves the rudder. The PCU achieves this thanks to two hydraulic systems working in tandem and pressurized by compressors driven by the engines, and by two separate electric motors.

The PCU is an ingenious hybrid, in more than one sense of the word. This single device combines the functions of two actuators, capable of pushing the rudder to the left and to the right. A device called the dual concentric servo valve, by moving just a minute fraction of an inch, diverts a high-pressure flow of hydraulic fluid down tiny passageways in a component the size of a soda can. The valve moves like a piston and is really two sliding valves (called "slides" by engineers), one inside the other, with passageways around the perimeter of each to allow the flow of pressurized fluid. When those passageways are lined up with one set of holes, or "ports," the rudder will move right; moved a few thousandths of an inch, they are lined up with another set of holes that moves the rudder left. If the inner, or primary, slide jams, the other—the outer secondary slide—can perform all the required functions on its own. Problems rarely occurred with this system. In fact, so few were expected that mechanics routinely ignored it except when a pilot reported rudder-control problems.

When the United mechanics dismantled Mack Moore's PCU, they discovered that one or both of the internal slides that comprise the dual concentric servo valve were traveling further than they should—all the way to an internal stop, in fact. When that occurred, there was significant leakage of hydraulic fluid from the part, and occasionally it jammed. But even more alarming, the actuator shaft, which the slides command, sometimes retracted instead of extending. Translated to rudder movement, that meant that the rudder would move in a direction opposite the one commanded by the pilot. Instead of going left, it would go right, and vice versa. By tweaking the external levers attached to the unit, the mechanics also discovered that they could reproduce the reversal even though the malfunction had otherwise disappeared. It appeared that overtravel of one of the two slides inside the unit was allowing hydraulic fluid to be routed through the wrong holes. Put simply, rudder commands were getting off at the wrong trolley stop.

Schleede reached Ed Soliday on the phone. He was United's director of corporate safety and the person ultimately responsible for the decision not to inform the NTSB about the intriguing possibilities emerging from the tail of Mack Moore's aircraft. Soliday admitted that he was familiar with the Mack Moore incident and had sanctioned the "undercover" investigation—a kind of "skunk works" operation. He told Schleede that he was frustrated by a number of his pilots and mechanics who claimed to have "solved Flight 585" when in fact they had done no such thing. Instead, when the Mack Moore incident came to his attention, he decided first to determine if it was related to Flight 585. Now the airline had decided that there was probably no linkage, and that the latest incident was an isolated anomaly.

This made Schleede see red, and he wrote a stinging rebuke to Dennis Lessard, United's senior flight safety investigator. "In other words," he charged, "UAL decided to gather the facts and analyze them before notifying the Safety Board. I don't approve of this method of cooperation during an open and unsolved investigation that could be remotely connected to such an incident."

Schleede had other reasons to be concerned about the Mack Moore incident. The NTSB had been called in by Panamanian authorities to try to explain why, on June 6, 1992, a Copa Airlines Boeing 737-200 had suddenly pitched into a steep descent and crashed in the jungle in Darien Province, killing all 47 people aboard. Preliminary results from the flight data recorder suggested the aircraft might have experienced flight anomalies similar to those of United 585. And the Air Line Pilots Association representatives on the Flight 585 investigation had raised several instances of rudders on other Boeing 737 aircraft behaving strangely as long ago as the 1970s.

On August 24, 1992, a group of senior NTSB investigators and their opposite numbers at Air Line Pilots Association, Boeing, United Airlines, and Parker Hannifin met at the Parker Hannifin plant in Irvine, California, to discuss the rudder PCU. The NTSB was represented by Al Dickinson, Merritt Birky, John Clark, John Delisi, and Greg Phillips. Phillips had just taken over the systems end of the investigation, and Birky was the NTSB's fire damage specialist. Also present was Wayne Gallimore, the machinists' union representative, who a year earlier had first raised the possibility that a defective PCU might be the cause of the Colorado Springs accident. United Airlines sent three people, Boeing four, and Parker Hannifin, the host, was represented by eight.

In the next several weeks, they would examine four Boeing 737 main power control units, all taken from planes involved in crashes or incidents. These were the unit from Mack Moore's airplane, the unit from Flight 585, another from a United aircraft that had made strange hissing noises during a fleet inspection, and the PCU extracted from the wreckage of the Copa plane that had crashed in Panama. The Parker Hannifin staff provided a test rig that pumped in pressurized fluid and simulated hydraulic conditions similar to those on board an aircraft. This enabled the investigators to put three of the PCUs completely through their paces, as though they were performing on a working aircraft. The fourth PCU, from Flight 585, was too badly damaged to be fully tested.

Boeing and Parker Hannifin personnel began by demonstrating how the PCU from the Mack Moore aircraft could misbehave. They admitted that Parker Hannifin's quality-control procedures would not have detected the defects in the unit that allowed the secondary slide, the outside of the two piston-like slides in the dual concentric servo valve, to travel past its normal stop and pump hydraulic fluid down the wrong holes, causing a rudder reversal. A new test procedure was adopted: pinning the primary and secondary slides together to see if the secondary slide could then be forced past its normal stop. The "hissing" PCU was also tested this way, and it, too, produced anomalies similar to those in the Mack Moore unit. The Copa unit demonstrated a pressure drop, suggesting rudder movement might be sluggish, but there was no reversal.

Although the 585 PCU was too badly damaged to be tested as a complete unit, the servo valve, which forms the heart of the component, was cleaned up and put through its paces. It demonstrated a pressure drop and reversal. However, Boeing personnel were quick to point out that the forces being applied to the unit were far greater than would be encountered in normal service. In an additional group of tests in late October, it was decided that no conditions existed where a complete rudder reversal was possible.

In mid-October, still puzzling over the anomalies they were unearthing, the NTSB investigators proposed rebuilding the entire 585 PCU, not just the servo valve, from scratch, using new parts where necessary to replace components damaged in the crash and subsequent fire. Then, when recreated to everybody's satisfaction, they wanted to hydraulically pressurize it to see if, as a complete unit, it would also duplicate the anomalies demonstrated by the Mack Moore unit.

These moves were energetically opposed by Boeing. John Purvis, Boeing's Director of Flight Safety and the man responsible for all of Boeing's accident investigation efforts, said reassembly could not restore pre-crash conditions because of the damage the unit had sustained, not just in the crash and subsequent fire, but also because of vigorous attempts to dismantle it (some of which involved the less than judicious application of a hammer). Not only would it be meaningless, but "hydraulically pressurizing a unit that has been subjected to severe impact, heat, and destructive disassembly is dangerous and Boeing does not recommend such an exercise." The idea was dropped.

A week previously, on October 8, 1992, a secret meeting of worried senior Boeing executives took place in the plane maker's Renton headquarters, outside Seattle. It was called by Jim Hutton, one of the engineers who had been given responsibility for solving the emerging problem of the 737 rudder, a problem that the Mack Moore incident and the tests at Parker had amplified. Nineteen Boeing executives, mostly engineers but also a legal specialist, were invited. (Nobody was invited from the NTSB or the FAA, nor were they informed about the meeting.) Their purpose was to arrive at a consensus for an action plan to present to the company vice-presidents. A PowerPoint presentation made to the meeting left nobody in doubt as to the gravity of the situation. The first slide read:

3 FUNDAMENTAL FACTS TO GET ACROSS TO VP'S

- *We have a problem*
- *We don't need to ground the fleet*
- *We do need a retrofit program to fix it*

SUPPORTING DATA

- *Valve doesn't meet fail-safe design intent*
- *Single valve jam gives potential for reversal*
- *Failure analysis says single valve jams only reduce max rate*

HAZARD ASSESSMENT

- *Loss of control judged to be on the order of* $10^{-8} \rightarrow 10^{-10}$
- *No known reversals in fleet history of* 5×10^{7} *hours*
- *This means we don't need to panic*
- *Does not mean we don't need to fix it*

A second set of slides was a classic cost/benefit analysis listing four possible options, ranging from doing nothing (unless a customer sought action) to an aggressive but costly replacement program. Safety was being measured not in terms of human lives, but in dollars and cents. Doing nothing exposed Boeing to the minimum cost but produced higher safety concerns and a certification issue. One interpretation of the latter implied that the aircraft might fail to measure up to the criteria applied to it by the FAA and, in a worst-case scenario, be grounded.

Next to doing nothing (the first option), the least costly option was to spread the retrofit over a 7-year or 15,000-flying-hour period; this second option reduced the safety concern somewhat. The third option, which appeared to be already finding favor among Boeing executives, was a 4-year replacement program that reduced safety concerns even further, but posed a higher cost to both Boeing and Parker Hannifin. Option four, a 2- to 3-year replacement program, posed the fewest safety concerns but, on the opposite side of the balance sheet, threatened "undue customer penalties" (in terms of service interruptions) and suffered from a "lack of parts." Option four, the presentation concluded, fixed the problem most quickly but posed the highest economic cost.

On November 5, 1992, Greg Phillips produced a full report on the PCU tests for the NTSB. On November 10, Carl Vogt, the NTSB's new chairman, wrote to Thomas Richards, his opposite number at the FAA, proposing that tests to detect PCU anomalies be developed by Boeing. Despite its mandate of promoting aviation safety, the NTSB cannot dictate terms to airlines and plane makers; it can only make recommendations about making rules to the FAA, the final arbiter of flight safety. Vogt's letter also mentioned work underway at Boeing and Parker Hannifin to redesign the PCU to limit overtravel of the secondary slide, and he added that Boeing was planning a retrofit program. His letter recommended that airlines be obliged to "incorporate design changes for the B-737 main rudder power control unit servo valve when these changes are made available by Boeing. These changes should preclude the possibility of rudder reversals attributed to overtravel of the secondary slide."

Although Vogt's letter can be read as conferring a moral obligation on Boeing to change the design of the PCU, nowhere does it suggest that Boeing be compelled to produce these changes or propose a deadline by which the redesign should be completed and the new units installed. Instead, in a rare departure from tradition, the NTSB appears to have reached a private agreement with a plane maker and one of its suppliers to rectify an airplane fault. Normally, the NTSB would have recommended to the FAA that Boeing be *required* to fix an offending part, and suggest a deadline for completion of the repair or retrofit. Now it was simply stating that a redesign was taking place and that airlines should install the part *when it was ready.*

There may have been benefits for both sides in this approach. Boeing and Parker Hannifin gained the image of being pro-actively concerned with airplane safety, of moving promptly to fix something without waiting to be instructed to do so by an arm of the federal government. Indeed, on a close reading of Vogt's letter, Boeing and Parker Hannifin are portrayed as partly instrumental in drawing the attention of authorities to the problem, not the other way around, and of initiating redesign of the offending part.

For its part, the NTSB may have been spared an agonizing wait to see how the FAA would react to any recommendation requiring action on the PCU from Boeing and Parker Hannifin. Thanks in no small measure to a provision in its original charter mandating the general promotion of aviation in the United States, the FAA saw itself not so much as aviation's policeman, but more as its partner, a sort of strict Federal Fairy Godmother. Many NTSB recommendations were rejected, or at least watered down, by the FAA, which cited their economic impact on airlines in return for what was seen as a tiny safety improvement. When safety advocates whined that the FAA had too cozy a relationship with airlines and plane makers, the FAA's customary response was twofold. One, we cannot regulate airlines out of business, and two, yes, we could implement these recommendations but then it will soon become too expensive for ordinary folk to buy a plane ticket.

Chairman Vogt could have been faced with the prospect of asking the FAA to ground a huge fleet in order to fix a condition that was extremely rare, something that presented minuscule danger to a slow-moving airplane with extended flaps taking off or landing, which had never been implicated in causing an accident, and which, if it did malfunction, could be overcome in almost all situations by an alert pilot. Of the five instances Vogt cited where a serious rudder malfunction occurred, none led to an accident, and four were traced to stray metal particles or corrosion in the PCU that may have caused a jam—more of a maintenance matter than an inherent design defect. The FAA might easily say, as it did so many times before and since, that the cost far outweighed the benefits and refuse to mandate the change.

Vogt was also getting ready for the long-delayed board meeting to decide the cause of Flight 585's crash. Under "sunshine" laws requiring federal agencies to conduct as much of their business as possible in public, the NTSB traditionally holds a public board meeting to discuss evidence unearthed during the course of an investigation. Interested parties—like an airline, a plane maker, or a labor union—can ask questions or challenge statements, but most of the meeting is pre-arranged to allow NTSB staffers to present the information they believe is most pertinent to the accident. This meeting is often followed, sometimes the next day, by another public board meeting at which the investigators' final conclusions are adopted. In reality, the board members will have seen and agreed upon the conclusions beforehand, after taking into consideration the views of the parties involved.

The case of Flight 585 was going to be different. A public meeting had already been postponed on the grounds that very little useful information was available. Now a public board meeting was going to be scheduled without a prior public disclosure of all the available evidence.

Not that the investigators hadn't tried their hardest. Apart from the various factual reports prepared by the NTSB's own expert groups, in June 1992, Al Dickinson had requested and received final submissions from the main parties to the investigation—Boeing, United Airlines, and the Air Line Pilots Association.

United's senior flight safety investigator, Dennis Lessard, had filed his submission under protest, saying there were still too many open items under investigation and that United regarded its submission as an interim one. This was backed by a letter from his superior, Ed Soliday, to Susan Coughlin, acting chairman of the NTSB prior to Vogt's appointment. "Our position is that offering final conclusions, recommendations, and probable-cause statements on this unique and technically complex accident before all investigative actions are complete and all factual data has been documented is premature and possibly counterproductive," Soliday wrote. "Indeed, such premature submissions may not only fail to address the real cause of the accident, but may even tend to obscure that real cause in a later analysis."

United's document was a masterful 45-page analysis of the evidence gathered to date. It offered no firm answers, although it nodded in the direction of rotors. "Based on the evidence available at this time, the cause of the abrupt event resulting in upset of the aircraft cannot be identified with a sufficient level of technical confidence. There is, however, sufficient historical precedent and evidence related to this accident to clearly link an unforeseen, violent weather phenomenon, most probably an extremely localized low altitude vortex or rotor, with the accident scenario."

The Air Lines Pilots Association (ALPA) was similarly hesitant to assign blame for the accident to any one cause. Its submission dismissed the rotor theory, saying that the wind profiles fed into the various flight simulator runs were based on making too many assumptions. "The concept of deriving winds and assuming a weather phenomenon to duplicate the flight dynamics is well intentioned but ALPA believes too much emphasis and reliance in the wind modeling has misdirected the investigation," ALPA's lead investigator, Captain D. B. Robinson, wrote. "Flight simulation is a valuable tool but the simulation is only as good as the theoretical assumed data loaded into the computer." Equally likely, Robinson's submission suggested, was a flight-control malfunction, possibly rudder- or aileron-related, and he could not rule out an autopilot defect either.

But Al Dickinson had long ruled out an autopilot defect. The cockpit voice recorder specialist Vincent Giuliana had spent many hours poring over the cockpit recordings, trying to detect anomalies in the engine sounds that might tell him whether or not one engine or the other was misbehaving. Some witnesses to the accident had said they thought they heard a popping noise. However, foreground noise in the cockpit toward the end of the recording stymied his efforts. He was luckier with an analysis of the clacking sounds made by the stabilizer trim wheels mounted to either side of the center console between the pilots. They spin at different speeds, depending on whether or not the stabilizer (which controls the aircraft's vertical orientation, i. e., nose up or down) is being operated by the autopilot or by the pilots via the control column. The noises he heard were the sounds produced by the slower spinning of the wheels caused by manual input, and that convinced him that the autopilot was switched off and could be eliminated as a suspect. Giuliana also failed to unearth any unusual, out-of-place bangs, creaks, or metallic groans that might indicate a flock of birds striking the plane or a break-up of the airplane.

ALPA pursued the rudder scenario like a dog worrying a bone. Even though the change in heading, as recorded on the flight data recorder, did not match the change expected from a rudder hardover, Captain Robinson said ALPA was "intuitively curious" as to what the effects of a hardover might be. He wanted to reproduce some of the possible rudder-control scenarios by flight-testing an aircraft. "The flight profile may well duplicate that of UAL 585," he added.

Dickinson and his team were not convinced that a rudder malfunction had occurred. Although the rudder was a prime suspect, there was no smoking gun. First of all, the condition of the PCU after it was salvaged from the wreckage effectively closed off an evidence trail. Not only was it seriously damaged by impact, it had been at the heart of a searing post-impact fire that caused further damage. Then, to cap everything, the unit had been damaged even further by efforts to pry it open in the lab, and some small components, including an internal spring, spring guide, and a cap from the servo valve, had mysteriously disappeared just as they were about to be dispatched to Parker Hannifin for further examination.

In addition, the way the aircraft behaved wasn't quite what could be deduced from a rudder malfunction. "The rate of the yaw and roll didn't seem to match anything we would expect from an airplane failure," explained Greg Phillips. "That's why we wondered if it might not be atmospheric. It was too slow. The calculations and simulations we had showed that the rate of the rudder going in would have been too slow. The roll was quick, but the rudder movement was too slow. We would have expected a failure to have been quicker than that."

There was less hesitation from Boeing about the probable cause, which the manufacturer blamed on "the airplane encounter with an unforecast and undetected high vorticity translating rotor near the ground in combination with a rapidly moving air mass which has been documented to have been in the area. This encounter at low altitude did not allow sufficient height for recovery."

Dickinson didn't quite know how to deal with the rotor issue. One of the things that stumped him was the failure to properly nail down exactly what the weather was doing at the time of the crash. Yes, there were rotors, but were they around Widefield Park at the time of the crash? He knew rotors descend from aloft; sometimes they remain 1,000 feet up, and sometimes they go right along the ground surface. But he couldn't get a precise measurement of how intense they were or how they might have affected 585's flight-control systems. The end result of the research indicated that the rotors were never very strong. He was now convinced you needed a rotor wind strength ten times greater than ever measured to affect the flight controls of that aircraft.

Nevertheless, the NTSB was gearing up to make the decision to blame the crash of 585 on the weather when the Mack Moore incident came to light. The subsequent investigation delayed things even further, making Flight 585 one of the longest-running investigations on the NTSB's books. Few investigations last more than a year, and several crashes that had occurred since had already been solved.

Al Dickinson had made several attempts to reinvigorate the investigation, pushing his group chairmen for more analysis, even debating whether or not a complete reconstruction of the wreckage would be worthwhile. Throughout the investigation several all-hands meetings were held in which, Dickinson said, they would "thrash out different theories and ask what else we should do and what else we should look at; that's how we function best as a party system, getting everybody's ideas so we never leave anything out. We went back out to United in California, where the systems were stored, and went through them again, a thorough examination of all the systems we had there, but it didn't lead to anything. I wish it did."

The Structures Group paid another visit to the Colorado Springs warehouse where the rest of the wreckage was still stored and sorted through it once again in a fruitless needle-in-a-haystack search for any evidence of problems that might have escaped their attention the previous year. The metallurgists in the

NTSB lab went over the ruined engines once more, patiently combing them for any evidence of pre-impact failure. The Cockpit Voice Recorder Group reconvened twice more to listen to the tapes, straining their ears to detect new sounds that might have escaped their attention during earlier sessions or to decide if anything they heard might benefit from a fresh interpretation.

Even a year after the wreckage had been cleared from Widefield Park, Tim Borson, Operations Group chairman, was back in Colorado Springs to see if he could pick up on what was a pretty flimsy lead. During a meeting back in the NTSB's Washington headquarters, a colleague had recalled a witness mentioning something about an elderly couple walking in the park. The elderly strollers had told the witness that a bad-smelling liquid fell on them from the doomed aircraft as it plummeted to the ground. This bit of information could be a vital clue, and perhaps it had been not so much overlooked as undervalued. United Airlines had provided the NTSB with samples of all the fluids routinely carried aboard its aircraft, including fuel, hydraulic fluid, engine oil, and even the blue chemical mixture used to flush lavatories. Borson, short on live leads, was determined to check it out. In the company of a colleague named Burt Simon, he flew to Colorado Springs for a week-long blitz of Widefield Park and its environs.

The two investigators arrived before 7 each morning and worked the park and the surrounding neighborhood until well after dark. They handed out hundreds of printed flyers containing a police sketch of what the elderly couple might look like. Everyone they saw in the park and in the surrounding streets got a flyer, as did every home in the neighborhood, and every store and restaurant got a stack. Borson and Simon walked and talked. They hung out. They knocked on doors. They even enlisted local radio and TV stations to broadcast descriptions and calls for help, and got the sheriff's office on a kind of standby alert. They wanted to be sure the whole town knew who they were looking for. But it was all to no avail. They found nothing about the elderly couple, or even the witness who'd initially described the couple. Even more discouraging, they encountered a whole new set of perfectly friendly and eager witnesses who volunteered that the weather on the fateful day was pleasant, with a light breeze.

Finally, on December 8, 1992, at a public board meeting, the NTSB officially declared that Flight 585 had crashed for "undetermined reasons," saying it could not identify conclusive evidence to explain the accident. No fault had been found with the crew, the airplane, the airline, or air traffic control, and the one faulty part found, the galled standby rudder actuator, would not have moved the rudder enough to push the plane into an uncontrollable dive. A rotor might have caused the plane to go out of control, but the flight data recorder readouts didn't support (or, for that matter, disprove) that theory.

However, Chairman Carl Vogt displayed an intriguing tendency to go with the rotor theory, and at one stage toward the conclusion of the meeting, he unexpectedly declared that the evidence appeared to favor a weather-related cause for the crash of Flight 585. A surprised Ron Schleede, the head of major investigations, was quickly on his feet arguing that the weather evidence was still inconclusive and raised the lack of corroborating evidence to support it. Vogt grinned at him and returned to the line that had been agreed on before the meeting. Afterward, Schleede challenged Vogt about his departure from the agreed procedure, and Vogt joked that he was just making sure that Schleede was paying attention. But in later conversations, Vogt revealed that he would have pushed the rotor theory had he received backing at the board meeting.

A lawyer in the transportation industry, Vogt came to White House attention when he was a board member of Amtrak, and this led to his appointment to NTSB by the first President Bush. However, he had also been an experienced navy pilot, flying F-8 Crusaders from the aircraft carrier *Lexington* on the eve of the Vietnam War. (He never saw action in that conflict, having finished his term of duty and returned to civilian life by the time it erupted.) Vogt's understanding of the rotor theory for Flight 585 suggested that the aircraft's flight could have been disrupted by a rotor without the low pressure core of the rotor registering as a false altitude gain on the aircraft's flight data recorder. The aircraft only needed to poke the tip of one wing into a strong rotor to be affected, Vogt reasoned, and this could happen without the altitude sensor, mounted on the forward fuselage, coming into contact with the rotor.

At any rate, the Board went with the "undetermined cause" decision and Vogt complied. This was a significant occasion. For only the fourth time in its history, the NTSB had failed to pronounce on a plane crash. The other events were the January 1969 crash of a Convair 440 in Bradford, Pennsylvania, that killed 11; the November 1970 crash of DC-9 in Huntington, West Virginia, that killed 75, including the Marshall University football team; and the March 1974 crash of another Convair that killed 36.

It had taken the NTSB 21 long months to reach its non-finding in the case of Flight 585, making it one of the longest investigations to date in its history.

Although it could find no reason for the crash of Flight 585, the NTSB had been concerned by what it learned about Boeing 737 rudder problems. In the period before announcing its non-finding in the crash, it issued a total of five Safety Recommendations related to the rudder. Two other Safety Recommendations called for broad research into the hazards of mountain weather and a special probe into the winds at Colorado Springs airport.

The first of these Safety Recommendations was issued on August 20, 1991, and related to the standby rudder mechanism where the NTSB had earlier detected galling. It asked the FAA to instruct airlines to check the force needed to rotate the input shaft on the standby rudder actuator and to see if the shaft bearing itself actually moved. If excessive force was needed or if the bearing rotated, the parts should be immediately removed from service and replaced with parts machined to finer tolerances.

In early October 1991, the FAA agreed and announced it was "considering the issuance of a notice of proposed rulemaking." In November, NTSB chairman James Kolstad said all was in order. The Safety Recommendation was classified as "Open—Acceptable Response," NTSB-speak meaning that the file would be closed after the FAA followed through. In February 1992, the FAA published a Notice of Proposed Rule Making (NPRM) in the *Federal Register*. Interested parties, usually airlines and pilots, are allowed two months to comment on the proposal. After assessing the comments, an Airworthiness Directive (AD) may be issued. Except in emergencies, the FAA rarely issues an AD or any change to existing procedures without going through the NPRM process. Its choice of the NPRM route showed that the FAA did not consider the rudder situation an emergency.

A month later, at the end of November 1991, the NTSB was making its displeasure known. Not only had the FAA's NPRM omitted the check on the bearing, it proposed allowing airlines 4,000 flight hours from the promulgation of the AD before they had to make the check on any aircraft. NTSB thought this was "excessive," as the check itself took only 6 hours. "Because the components affected could cause an uncommanded rudder input, the Safety Board believes that these inspections should be performed as soon as possible or at the very least at the next available inspection of the airplane," wrote Susan Coughlin to her opposite number at the FAA. (Coughlin had become acting chairman following Kolstad's departure at the end of his term of office.)

There was even more alarm at the NTSB when, on August 5, 1993, the FAA announced that it had changed its mind and was withdrawing the NPRM. The FAA had reviewed the 737 rudder system's design and decided that there was absolutely no reason to make changes. It had decided that in pre-flight checks, pilots easily detected increased resistance in the rudder pedals caused by jamming, through galling, of the standby actuator shaft. In addition, the FAA believed, there would also be easily detected "erratic nose wheel steering with the yaw damper engaged" as well as "kick-backs" felt on the rudder pedals during flight, and the yaw damper would produce "erratic operations."

"The FAA has determined that the condition addressed in the NPRM is not an unsafe condition warranting the issuance of an Airworthiness Directive," the agency's administrator wrote to NTSB Chairman Vogt.

In addition, the FAA decided that the 737 control system linkages were flexible enough for a rudder command to reach the input lever of the main PCU even if the standby input shaft were completely welded to the bearing. The FAA's official statement explained that "full rudder can be compensated with lateral controls in the majority of flight envelopes." In other words, except at low speeds, the pilots could counteract a jammed rudder by using the hinged surfaces on the wings called ailerons and spoilers. And finally, "Boeing has revised the Model 737 maintenance manuals to emphasize the indications of input lever binding in the standby rudder PCU, which would facilitate an operator's ability to determine the proper maintenance action."

Carl Vogt was not very impressed with this chain of thought but gave in to the FAA's reasoning and decision. On November 15, 1993, he replied that he was pleased to learn that galling was detectable by the pilot and was not an unsafe condition. However, he added, it could result in erratic flight control that could distract a flight crew and be potentially hazardous. Given that there was no further evidence that it could cause significant uncommanded rudder deflections, he was closing the file on the grounds that "Acceptable Alternative Action" had been taken.

On January 19, 1993, the FAA responded to the other Safety Recommendations made by the NTSB prior to its final report on Flight 585. Those Safety Recommendations had asked that an examination be made to detect faulty PCUs, and that they be changed as soon as new parts were designed and issued. The FAA said that Boeing was issuing service information to airlines to inspect and retrofit 737s as the parts became available. At that point the FAA would consider issuing an NPRM. The NTSB replied that this was an acceptable response, although the file was marked "Open" pending a final resolution of the issue. The FAA also declared it was going to publish an NPRM requiring several special maintenance checks on PCUs. On the issue of developing a pilot's pre-flight check, the agency said that the normal checks were adequate. But the NTSB was worried that the rapid pressing of the pedals needed to provoke a servo valve jam could damage the aircraft, and Vogt requested the FAA to reconsider its response. The file for this request was marked "Open—Unacceptable Response." On the issue of checking all other servo valves manufactured by Parker Hannifin, the FAA reported that the 737 PCU was the only one causing trouble; therefore, Vogt marked the file on this "Closed—Acceptable Response".

In November 1993, the FAA finally issued its NPRM about checking the main PCU, an issue the NTSB had raised in the wake of the Mack Moore incident. It proposed requiring specially trained operators to perform periodic inspections at intervals of not more than 750 flight hours until the new valves came from Boeing. During inspection, the rudder pedals would be cycled at the maximum rate, and special instruments and additional observers would be used to properly detect any anomalies. Airlines would have five years to modify the servo valve from the effective date of the new rule.

The inspection procedure, the FAA's administrator David Hinson wrote to Vogt on December 2, 1993, was specially developed by Boeing to identify excessive internal leakage in the main rudder power control unit servo valve, which is a symptom of secondary slide overtravel. "The FAA believes that the combination of routine preflight checks and a dedicated, periodic ground test offer the best overall method to ensure proper rudder operation."

The FAA's NPRM stated that there were 2,448 aircraft with the affected design in the world fleet, including 729 in the US. The NPRM had no qualms about doing the math. It would take 19 hours per airplane at $55 per hour to accomplish the proposed test actions. That added up to $1,045 per aircraft, a total of $761,805 for the US fleet.

Several airlines (and pilots organizations) raised objections to the NPRM in letters to the FAA. Many of them wanted a single, one-time only test, not a repetitive test after every 750 hours of flight. In any event, they said, the proposed test was unreliable, and decisions to replace what might prove to be perfectly serviceable (and expensive) PCUs could be made on a faulty judgement call by mechanics. Others wanted the compliance period for modifying the servo valve pushed even further away, from five years to seven years. A single commentator wanted it shortened because aircraft could otherwise be operating with defective units for up to five years. The FAA disagreed: "The repetitive tests required by this AD [sic] will provide an acceptable level of safety in the interim." It allowed one change only, in response to a commentator's request for replacement of a faulty unit by a serviceable, but unmodified, PCU, providing this was replaced with a PCU of the new design at the end of the five-year period.

On January 3, 1994, the FAA issued the first AD based on the previous November's NPRM. It required the development of a test of the main PCU and the replacement of the PCUs when new ones became available. A second AD, issued in July 1994, required a PCU inspection every 750 hours of flight time with equipment that Boeing had designed. On August 11, the new NTSB acting chairman Jim Hall wrote that the safety recommendation files were closed due to acceptable action on the part of the FAA: "The Safety Board believes that in the interest of safety all Boeing 737 main rudder PCUs should be modified at the earliest possible date, and since the compliance period appears to be founded on reasonable estimates of equipment availability, the Safety Board classifies these recommendations Closed—Acceptable Action." Availability, he agreed, hinged on the ability of Parker Hannifin to produce the hand-made servo valve parts.

Airlines now had the task of implementing the two Airworthiness Directives and the service bulletin issued earlier by Boeing. But one airline, USAir, had been spurred into early action when one of its Boeing 737 aircraft suddenly displayed "Mack Moore" symptoms early in 1993. The company alerted pilots and crews to the PCU issues and said that every 737 in its fleet had recently been tested even though "Boeing has not recommended any operator action or testing to date." All its aircraft passed the test, USAir reassuringly declared. Pilots were advised to conduct the usual preflight checks on the flight controls. "It should be obvious that flight must not be attempted if any controls fail to exhibit unrestricted movement."

In July 1993, Boeing issued a Flight Operations Review document highlighting the actions pilots should take if they experienced "in-flight events which are beyond the scope of established non-normal procedures." This could be a mid-air collision, a bomb explosion, or "other major malfunction." Reassuringly, it stated: "If a jammed flight controls condition exists, both pilots can apply force to either clear the jam or activate the break-out feature designed into all Boeing airplanes. There should be no concern about damaging the mechanism by applying too much force."

Boeing had published its long awaited rudder PCU Service Bulletin on April 15, 1993, and revised it a year later. Part of the revised bulletin read: "Boeing recommends that the operator do the changes given in this service bulletin as soon as parts are available and the rudder power control unit is removed for scheduled or unscheduled maintenance. Boeing also recommends that the operators do the change on their spare rudder PCUs as soon as parts are available. There is a limited number of parts available from the supplier."

Overall, however, there was nothing to alarm airlines or pilots, and the vast majority of passengers were unaware that anything about the USA's most popular aircraft was in any way open to question. But less than one month after the NTSB's new acting chairman, Jim Hall, finally closed the file on the various Safety Recommendations that had emerged from the investigation of Flight 585, the unthinkable happened.

Part Two

427

hapter 6 OUT OF THE SKY

On Thursday evening, September 8, 1994, Amy Giza was waiting with her five-year-old son David in her parked car in the rural outskirts of Aliquippa, a few miles northwest of Pittsburgh. She was reading a book and trying to ignore the constant distractions provided by David. Shortly after 7 o'clock, when a plane went overhead with a noise loud enough to be irritating, she assumed it was a small plane flying very low, and she did not look up. Then the noise changed to a sort of whine like she had heard at fireworks displays when big rockets go up. David said, "Mommy, that plane just fell out of the sky." She looked up but saw nothing. "What kind of plane?" she asked. "A giant plane," he replied. Then the noise stopped. Amy Giza didn't know it, but another 737 had inexplicably fallen from the skies.

Because Aliquippa lies below a busy main approach to Pittsburgh International airport, with aircraft passing overhead every two or three minutes, most local people have grown accustomed to the noise, hardly giving a thought to the overhead roars. But there was something about the whine of Flight 427's engines that made dozens of people look up, almost simultaneously, and see its death dive.

Robert Cellini was at the car dealership where he was a sales manager when a colleague commented on a loud noise, as if an aircraft were powering up its engines. Cellini went to the door, opened it, and saw in the distance a dark-colored airplane with a USAir emblem on the tail, its nose pointing toward the ground "like a bullet," giving off a screaming whine. At the moment the plane disappeared behind some trees, Cellini instinctively closed the door. There was an explosion and the door shook. Through the glass he saw a red and black ball of flames and smoke billowing up beyond the nearby woods.

In the same showroom, Carl Anderson ducked under a desk as the noise kept getting louder, as if the plane were coming right at their building. While Cellini went to the door, Anderson watched, horrified, through a window. He saw what at first he took to be a jet fighter heading north. It seemed as if the pilot was trying to pull up the nose and turn at the same time. Anderson had the impression that the pilot, while in trouble, was trying to steer clear of the nearby nuclear power plant and the mall where the showroom was located. "This guy is a hero," he thought, as he saw the airplane bank, then pitch nose down and out of sight, like an incoming rocket. Anderson dialed 911 and told the operator what he had seen, then ran up to a nearby intersection of the highway to Aliquippa to see if he could help by directing fire engines and ambulances.

Cellini, meanwhile, jumped in a car with two other colleagues and drove the mile or so to the crash site, a ravine at the side of a dirt road. They ran about yelling to see if anyone needed help. Apart from the roar of the flames and the crackling of burning vegetation, they heard nothing.

Jeri Beam was at a nearby soccer field watching a kids' game when someone yelled, "Hey look!" and pointed up. Beam saw what appeared to be the top side of a jetliner as it flew low overhead, completely upside down, with the nose pointed slightly toward the ground. Then, nose first, it dived out of sight into some trees, and a giant fireball erupted.

There were other witnesses at the same soccer field. Mark Barney, a 35-year-old sales representative and coach, glanced up and saw a jetliner aimed like a dive-bomber at the trees. To Becky Meuser, it appeared as if the plane wobbled slightly, then was "corkscrewing" down to the point of impact. Denise Clark witnessed the crash from outside her home on a hilltop about three miles away. Her son, who had seen many aircraft pass overhead, remarked, "Mom, that one's real low." She watched as it appeared to turn to the left. As it started to bank, it's right wing suddenly lifted until it was perpendicular to the ground. Then the nose gradually dropped, and it disappeared behind a hill.

Brenda Saakis was in the parking lot at the nearby Hills Plaza mall when she saw the low-flying plane. For a moment its wings were level, but then it suddenly banked right, veered to the left, turned upside down, and disappeared nose down. The plane was too large to be doing aerobatics, she thought.

As is always the case, different people saw different things. One man said an engine was twisted on a wing; others said both engines appeared normal. Several people said they saw smoke pouring from an engine, a wing, or part of the fuselage. Others saw no such thing. Some said the whining sound of the engine was progressively louder and higher-pitched, others said it was intermittent, punctuated by fits and starts, as if the pilots were trying to restart from a stall. One woman said it sounded like an old car trying to start on a cold day. Some said the plane corkscrewed, others said it didn't. Some said it banked right, others said it banked left. Like Carl Anderson, several witnesses were convinced the pilot deliberately crashed the plane in the ravine to avoid populated areas.

One of the first NTSB investigators to reach the scene was Chuck Leonard, a former Eastern Airlines captain. He had worked as a senior operations investigator for the NTSB since 1989, when he found himself out of a job after Eastern was taken over by corporate raider Carl Icahn. Because he was based at the NTSB's New Jersey regional office, Leonard was able to board a scheduled flight and reach the scene before the first contingent from Washington was even off the ground. He was met at the airport by a police escort and taken, blue lights flashing, to the emergency services command post set up at the accident location. Before he had a chance to examine the crash scene, he was introduced to two FBI agents who took him to one side and told him that an FBI informant who was helping prepare testimony for a drug trial in Chicago had been aboard the aircraft. Already, information coming in from witnesses suggested that smoke might have come out of the fuselage, not far from the baggage hold. The FBI, Leonard was told in confidence, would be treating this accident as a possible murder and would be seeking the NTSB's help in obtaining relevant parts of the aircraft to test for explosives in the FBI's laboratories.

Leonard had a brief glimpse of the accident scene, now lit by arc lamps as preparations were made for the following day's grim task of recovering the remains of the 132 people who had been aboard. Later that night he was joined by the Go Team from Washington, which was led by Tom Hauteur, the NTSB's investigator-in-charge and Leonard's supervisor. Hauteur's youthful appearance belied the fact that he was a ten-year veteran of the NTSB with impressive credentials and wide-ranging experience. He had qualified as a private pilot at just 16 years of age and, after graduating from Purdue with a degree in aeronautical and astronautical engineering, took a master's in business administration from George Mason University. Right out of graduate school, he was hired by United Technologies, a top high-tech conglomerate whose holdings include Pratt & Whitney, maker of engines that are fitted to many US jetliners. Hauteur worked on their JT-9 series of engines. Then he worked for a Washington aeronautical engineering consulting firm that was designing new advanced plastic composite wings for the US Marine Corps Harrier Jump Jet. And when that contract assignment was completed, he went to work for Tracor Corporation, which was looking for ways to generate electricity from the temperature differences between seawater at the surface and at various depths. It was an extremely exciting and groundbreaking notion, but it turned out to be impractical. Tracor dropped the idea, and Hauteur found himself managing the firm's R&D division for a few years. When Tracor fell on hard times, Hauteur found himself looking for a new job and landed one at the NTSB.

At the NTSB, Hauteur's previous involvement with the Boeing 737 had been in 1992, helping Panamanian authorities try to fathom why a Copa Airlines plane crashed into the jungle in Darien Province. At first, it had been thought there were significant similarities between the Copa crash and the United Flight 585 accident in Colorado Springs, but eventually the two investigations took different directions. A frayed short-circuiting wire was discovered in the Copa aircraft's autogyro, the instrument that feeds an aircraft's heading into the attitude indicator, which, in turn, tells a pilot whether his plane is banked or level, and whether he is climbing or descending.

Take away the usual reference points a pilot depends upon—that the sky should always be above, the earth below, and the horizon parallel to his wings—and the pilot is helpless. The attitude indicator is a lifesaver because without looking out the window, a pilot can get feedback on these three key parameters. But if he relies on an attitude indicator when it is faulty, the device becomes a signpost to disaster. Hand-flying the aircraft at night without any visible horizon for a benchmark, the Copa pilots had been attempting to maintain course by using instruments. However, not realizing a short circuit was causing the attitude indicator to give wildly misleading readings, they had actually turned the aircraft on its back, and from there it went into an uncontrollable dive and broke up in midair before crashing with the loss of all 47 aboard. It was the similarity between the paths traced through the air by the two 737s in their final moments that had attracted the attention of Al Dickinson's team probing United Flight 585. But the discovery of the Copa plane's defective attitude indicator put a quick end to that line of investigation.

Hauteur was aware of these issues surrounding the 737, and he was experienced at managing crash investigations, having worked on several high-profile accidents as investigator-in-charge. But it had been six years since he had taken full control of a disaster investigation on this scale. In August 1988, he had supervised the investigation into Delta Flight 1141, a Boeing 727 that crashed in Dallas leaving 14 dead and 76 injured, 26 seriously. After failing to gain altitude on taking off, the plane had collided with a navigational antenna 1,000 feet past the end of the runway. The captain, who survived, said it felt like the aircraft was experiencing reverse thrust on takeoff. But careful forensic work by Hauteur's team determined that, because of slack cockpit discipline, the pilots had attempted to take off without deploying the proper wing flaps and slats. A warning system, which should have alerted them to their error, failed to work. And Delta's operating procedures, including crew training, were found wanting during a recent period of rapid corporate growth. So was the FAA's supervision of the airline, given that it was aware of Delta's training and discipline problems.

Since then, Hauteur had worked on two other high-profile crashes, but both involved considerably smaller planes. In April 1991, he headed the investigation into the Brunswick, Georgia, crash of an Atlantic Southeast Airlines Embraer 120RT that killed 23, including former Senator John Tower. Here, the culprit was a gear malfunction that caused one propeller to suddenly go into idle mode. Hauteur's investigation led to urgent maintenance action on the entire Embraer turboprop fleet and all other aircraft using the same engines. Also in April 1991, he was appointed to lead an investigation in Merion, Pennsylvania, where there had been a midair collision of a private aircraft and a helicopter. In that accident, US Senator John Heinz and six others, including two schoolchildren on the ground, were killed. It emerged from the investigation that the collision occurred after the helicopter elected to fly close to the Piper Aerostar, in which Senator Heinz was a passenger, to report on whether the nosewheel was properly extended for landing. Hauteur's report blamed both pilots for "poor judgement."

But 427 was of a different order of magnitude, and in taking charge of the meetings held on the day after the crash, Hauteur wanted to clearly establish the lines of authority, and clearly establish who was running things before they got out of hand. One of his first acts, in a Holiday Inn meeting room near the crash site, was to ask a USAir vice-president to leave the room because the executive wanted to act solely as an "observer" on behalf of his company. (He did not want to be assigned to any NTSB-led team.) Hauteur's polite yet brutal frankness in asking him to leave left the airline executive speechless. When he recovered his composure, he protested, saying his company's rules obliged him to remain and note what was happening. But Hauteur was unrelenting; this was his meeting and his investigation and he would decide who remained and who left. Hauteur was following strict NTSB guidelines in moving quickly to establish "ownership" of the investigation. There were too many outside reputations at stake after any accident, and he feared he and his staff could be put under intolerable pressure if he allowed a posse of corporate suits to lurk around the command center.

Hauteur had started by introducing his chairmen, including Chuck Leonard, who as head of the Operations Group would be investigating the history of the flight, the pilots, and the operating and training environment within USAir. Leonard had earlier informed Hauteur of the FBI's interest in the crash, but Hauteur decided to say nothing for the moment, apart from briefing his colleague Cynthia Keegan, chair of the Structures Group. Keegan was faced with the mammoth task of managing the crash site in addition to collecting and identifying all the wreckage. She had already noticed FBI agents in the room, but that was not unusual at the early stages of a plane crash investigation. In addition, they were offering the services of their forensic unit to help the coroner's staff identify the dead, an offer that was gratefully accepted. Others in the room included Jerome Frechette, one of the NTSB's team of aircraft engine experts who was to head up the Powerplants Group. His function would be to determine what role, if any, the engines may have played in the disaster. Also taking a seat was Al Lebo, a former air traffic controller, now responsible for finding out how the Pittsburgh tower handled the flight and asking if they could have done things differently. As he moved his gaze around the room, Hauteur recognized more familiar faces. Beverly Johnson was there to coordinate the hundreds of statements likely to be taken from witnesses. Lending her a hand was Malcolm Brenner, a psychologist who would chair a Human Performance group when the facts of the accident were more established. He would ask, among other things, if too much was expected of the crew of the doomed airplane. James Skeen, one of the NTSB's meteorologists, would try to discover if the weather had any role in precipitating the crash of Flight 427. Tom Jacky would attempt, by analyzing flight data, to reconstruct the aircraft's last moments in the air. Another colleague, Greg Philllips, would probe the aircraft's electrical and hydraulic systems to see if they had been faulty. Henry Hughes was destined to chair one of the shortest-lived probes in the whole investigation. His job as Survival Factors chair was to discover what contributed to the escape of survivors. Hector Casanova was to leaf through USAir's maintenance records to discover if flight 427 had been looked after properly.

Some of the staff knew a great deal about the 737: Lebo, Brenner, Jacky, and Phillips had all been directly involved with the Flight 585 probe. Others had varying degrees of familiarity with the plane and its "issues." And two key NTSB group chairmen—Albert Reitan, the cockpit voice recorder specialist, and Jeremy Akel, the flight data recorder specialist—were not even present, but back in Washington awaiting delivery of the black boxes from the crash scene. But the whole NTSB team, Hauteur insisted, had to embark on this investigation with the most scrupulous attention to and respect for details, regardless of where they might lead the investigation.

Hauteur also, of course, asked representatives of the various outside parties to introduce themselves. Among others, they included people from Boeing, the Federal Aviation Administration, USAir, Parker Hannifin (manufacturer of many of the aircraft's hydraulics systems), the Air Line Pilots Association (two of whose members had died in the crash), the International Association of Machinists and Aerospace Workers (representing aircraft mechanics), and the Association of Flight Attendants. Some of them had worked on previous accidents and some had even taken courses in air accident investigation, such as those offered by the University of Southern California. Some expressed an interest in joining particular groups, and it was usually fairly self-evident where they would be assigned. For example, an employee of CFM International, which manufactured Flight 427's engines, was going nowhere other than the Powerplants Group. Sometimes a party representative's request to be put on a particular group didn't work out. For example, Robert Sumwalt, a USAir captain and a delegate from the Air Line Pilots Association, asked to be put on the Human Performance Group. Because this group had not yet been constituted, he ended up collecting witness statements instead. Almost without exception, every group had a representative from the FAA, the pilots' union, and from Boeing, whose employees were packed into the room. Of the 24 people who would eventually make up Cynthia Keegan's Structures Group, 11 were from Boeing.

The next step for Hauteur and his investigators was a tour of the crash scene. "Look, don't touch," Hauteur cautioned those who accompanied him on the first walkabout. Despite the fire and despite the efforts of the emergency and law enforcement teams the night before, the wreckage was still strewn with human remains. Chuck Leonard was horrified that the only piece of human anatomy he could recognize was the arm of one of the cockpit crew emerging from the wreckage, identifiable only by fragments of his uniform. Leaving the remains where they lay overnight may have appeared callous, but it had a humane purpose. Given the state of dismemberment of the bodies, the coroner's staff might stand a better chance of identifying them if they could be located according to seat numbers. This, it would turn out, was a forlorn hope.

Flight 427 hit the ground at a speed of some 300 mph. While airplane occupants have survived high-speed crashes, it is usually only because the plane comes to earth at a shallow angle, then slides along the ground for some distance, dissipating energy as it goes. Its descent almost couldn't have been steeper, and the occupants, to put it bluntly, never had a chance. The aircraft hit upward-sloping ground so hard it exploded on impact, showering debris for hundreds of yards. Some parts of the plane, including one of the tires, were not located for almost two months afterward.

Looking at the ominous scene, Keegan was close to despair. How would she make sense of it all? "Where is everything?" she later reported as her first reaction. Apart from the remains of the tailplane, there were no recognizable sections of aircraft, and it was some time before she realized that pieces of wreckage were strewn over thousands of square yards. The night before, when she'd gotten the call to drop everything and join the Go Team at Washington's National Airport, Keegan had been getting ready to meet her mother and some friends for a B.B. King concert. She tuned in to one of the early TV news bulletins on the crash as she packed a bag. From the outset the reports were saying there were no survivors. After two years of working mostly on helicopter, general aviation, and commuter accidents, she realized this was going to be very different. This was a major accident.

A native of Lebanon, Pennsylvania, Keegan graduated in aeronautical engineering from Embry-Riddle Aeronautical University in Florida, majoring in aircraft maintenance. She also earned her pilot's license. In 1983, she landed a two-year stint with the NTSB as part of its cooperative education program and received basic training in aircraft accident investigation. For the following seven years she worked as an engineer for McDonnell Douglas on aircraft as diverse as the MD-80, the DC-9, the DC-8, and the Apache helicopter. But getting back to the NTSB had been her ambition, and with large airplane experience under her belt, she applied and was hired in September 1992.

Now, tramping around the devastated hillside in Pennsylvania, she knew that it would fall to her Structures Group to make sense of the bewildering puzzle of fragments strewn across the ground. This was the first time she had responsibility for such a large crash scene, and the prospect was daunting. Already, she could see that the crash had attracted major media and political attention. It was likely, therefore, that Hauteur would be pinned down at the command center dealing with administrative issues and would be unable to offer her much support in managing the crash scene. The glare of publicity, too, was something new to her.

Once upon a time, structural failures were a significant cause of aircraft accidents. Wings snapped, engines and doors fell off, fuselages split open. But with tougher regulations and improved manufacturing and maintenance techniques, structural failures began to fade in significance—or at least in frequency. In American airspace, the last major airline accident directly attributable to structural failure was the remarkable Aloha Airlines incident in April 1988. In this startling and nearly disastrous accident, an entire 18-foot section of the cabin wall and roof structure of an elderly and haphazardly maintained Aloha Boeing 737 blew off into the Pacific, leading to the death of a flight attendant and exposing six rows of passengers to the elements—at an altitude of 24,000 feet and a speed of more than 300 mph. Amazingly, the pilots managed to land the crippled airliner safely on the Hawaiian island of Maui, but not before several passengers lost consciousness from oxygen starvation and the cold. The airplane's nose was drooping more than three feet out of alignment after the landing. This had happened because the critically weakened fuselage almost snapped in two during the perilous descent, the last part of which was conducted on just one engine.

The Aloha incident was an anomaly, found to be caused by inadequately repaired corrosion damage. If anything, structural failure was the least likely of all causes of a modern airliner crash. This fact didn't, though, make the work of the Structures Group any less important. In fact, in the early stages of an accident investigation, it was the single most important group at the scene. For one thing, Structures was in charge of the crash site and debris. Just collecting what remained of the plane, and keeping this debris in an orderly and comprehensible order, presents a huge challenge. Moreover, since the investigator-in-charge is usually tied down at the command center dealing with administrative issues, direct on-site management of the teams often devolves to the Structures Group chair. Keegan knew that in the case of Flight 427 she would be expected to take matters in hand on the hillside.

But first she had to make sure that she could account for and identify as much of the wreckage as possible. She was not alone in wondering if sense could be made from Flight 427's mangled and shredded parts. Her Structures Group colleague Frank Hildrup was equally aware that mass destruction, on the scale he was now witnessing, could make the job extremely difficult. His mind went back to Colorado Springs, less than four years earlier, where investigators faced similar devastation. Some highly fragmented airplane parts were so deeply buried in the hole that Flight 585 had dug for itself that the only solution had been to excavate with a backhoe, then sift the soil and wreckage mixture using a mechanized sieve similar to those used in quarries for sorting gravel. (Using such a "fine-toothed comb" was a highly unusual practice and indicated how little tangible evidence had been recovered and identified.)

And it wasn't just hard evidence that was in short supply. Apart from FBI suspicions of a bomb on board, the investigators had no idea what direction their investigation would take. They had no prevailing plausible theory. Except for a garbled emergency call to the Pittsburgh air traffic controllers, there had been no indication from the cockpit that anything was wrong. And the fewer the clues, the more likely that a reconstruction of the wreckage might need to be attempted, if only to rule out structural factors or the failure of key control surfaces like ailerons, elevators, and the rudder.

Keegan assembled her group at the crash site the day after the plane went down, immediately following Tom Hauteur's tour of the wreckage. For the next seven harrowing days the bloodied, scorched hillside and woods of the ravine were her team's workplace, the smell of death and aviation fuel constantly in their nostrils. Because the coroner's staff needed to disturb wreckage in the search for bodies and body parts, the Structures people had to work alongside them, with notebooks and cameras at the ready. As much as possible, they had to document exactly how the wreckage had been left by the crash, before it was disturbed by other investigators. (The Structures people also helped the coroner's staff by identifying the portions of the airplane in which human remains were found.)

Seeing almost unrecognizable human heads, bodies, limbs, and organs, and wreckage smeared with blood was bad enough. Even more upsetting was watching the very unpleasant work the coroner's staff had to do, and not everyone could take it. Every now and then someone would walk purposefully a few yards into the woods and lean against a tree with his or her back turned to colleagues and the wreckage. Sometimes you could tell someone was sobbing by the heaving of their shoulders, sometimes not.

To the emotional trauma was added the physical discomfort of working inside biohazard suits. Because of the large amount of blood around the site, it was decided that crash investigators and coroner's staff must wear these suits as protection against possible infection from blood-borne pathogens, notably hepatitis and HIV. In the early autumn heat, the suits were extremely uncomfortable; going to the bathroom or taking a break was a long, drawn-out affair requiring the wearer to go through a decontamination station.

Apart from noting the position and condition of wreckage about to be disturbed by the recovery of bodies, Keegan had to document where other pieces of wreckage had fallen and attempt to decide how it had got there. She and her staff had divided the crash site into quadrants and started noting the location of each piece recovered in the minutest detail. Apart from the tailplane and a large part of the left wing, there were no immediately identifiable major parts of the airplane at or near the main wreckage site. The left wing had been the first part of the airplane to come to earth, leaving a 25-foot-long scar on the top of a rise overlooking the main crash scene before snapping off and coming to rest midway between the scar and the main wreckage. It rested on a burning engine, and the fire had softened the wing until it was draped over the engine like an old tarpaulin, not the rigid airfoil it had once been. The bulk of the additional wreckage was strewn in a 350-foot radius around the main impact crater. Some pieces of wreckage were buried so deeply they were not immediately evident, and metal detectors and ground-penetrating radar equipment loaned by the US Bureau of Mines were used to find pieces of wreckage that had ploughed deep into soil. Some pieces were later recovered from 8 feet underground.

Apart from the scar from the wing and the main impact crater, the landscape had an additional story to tell. Early on, Keegan detailed investigators to look for fresh marks in the ground, broken branches, and snapped-off treetops. Careful measurements were made of broken tree trunks and branches to determine the angle at which they had been struck by wreckage. Vertical breaks of branches on a tree beside the scar made by the wing suggested that the wing was in an almost vertical position when it struck the tree, evidence that the aircraft had been on its side when the wing struck the ground. The trail of tree damage prior to the scar gave a further clue as to the direction and angle at which the aircraft fell to earth. It was clear was that the plane had not "pancaked" onto the hillside, but come in at a steep angle.

The location of scorched and shriveled vegetation was mapped on the very first day of the investigation. As the week wore on, a careful watch was made to see if additional trees and vegetation displayed symptoms. Shriveled or discolored leaves appearing several days after a crash could be evidence of fuel contamination. If the tree is under the final flight path, but some distance back from where the wing fuel tanks came to rest, the question must be asked: was the aircraft leaking fuel as it came down, and did that have a bearing on the accident? Or were the trees simply showered with atomized fuel as the fuel tanks burst? An L-shaped area around the main impact site contained badly burned and broken vegetation. North of that was a 40-foot-wide, 175-foot-long swath of vegetation discolored by fuel with two similarly discolored patches to the south. There was no obvious pattern to the discoloration, which had evidently been caused by the fuel tanks bursting on impact and haphazardly showering fuel around the immediate neighborhood.

The discovery of the flight data recorder and the cockpit voice recorder early in the investigation came as a relief to many on the scene, but more seasoned hands like Keegan were cautious. They knew better. The flight data recorder would say what the plane did, not why it did it. And there was only a slim chance that the crew had verbally exchanged any more information in the cockpit than they'd imparted to the air traffic controller. When word filtered back from Washington that the initial cockpit voice recorder analysis appeared to have drawn a blank in terms of pointing to a cause, Keegan realized that more of the subsequent investigation was going to depend on her. And if no more clues emerged over the coming days, there was a strong likelihood that Hauteur would order a reconstruction of the wreckage.

But before that task could be undertaken, even before a decision to reconstruct was reached, Keegan had another important job to do. Armed with a computerized model of Flight 427's last moments, she directed a helicopter and a team of volunteers to a large area near the soccer field over which the doomed aircraft had passed shortly before crashing. Their mission was to uncover any objects that may have fallen off the airplane prior to its tumbling out of the sky. At least two witnesses had said they thought they saw something coming off the aircraft before the impact; one was a six-year-old child. An adult witness said one of the engines appeared twisted on the wing. However, the volunteers came up empty-handed. Nothing was found to have fallen before the crash.

The crash scene was now secure. It was time for Flight 427 to complete its interrupted journey to Pittsburgh. This time, it was going on the backs of a fleet of trucks.

Chapter 7 A TARGET HAS DISAPPEARED

Chuck Leonard's Operations Group had a very broad mission. Not only was it charged with finding out how USAir Flight 427 was piloted on its doomed flight, it had to probe, in the minutest detail, the background of the crew and the airline's procedures and practices. Leonard wanted to know, for example, how USAir recruited and trained its pilots, how their performance was scrutinized, and how they were trained to handle emergencies. Was there anything about the safety culture at USAir that might explain this and any other accidents? Was the aircraft fueled and loaded properly? Had the aircraft, in its maintenance schedule, shown any recurring or unrectified problems? Were the pilots fatigued?

The day of the crash was the third in succession that Captain Peter Germano and his co-pilot, First Officer Charles B. Emmett, had worked together. On the first day, they flew from Philadelphia to Indianapolis, back to Philadelphia, and then to Toronto, where they spent the night. On the second day, they flew from Toronto to Philadelphia, then Cleveland, then Charlotte, and then, shortly before 11 P.M., they arrived at Jacksonville, where they would spend their last night alive. While they slept, a Boeing 737 with tail number N513AU—the aircraft they would use the following day—was being prepared in Windsor Locks, Connecticut.

It was a Boeing 737-300, a more up-to-date version than the 737-200 that had crashed some three years earlier in Colorado Springs. The most distinguishing feature of this newer model was its engines. They were larger, fatter, and positioned farther forward on the wing of the aircraft by comparison to the slimmer, cigar-like engines slung below the wing of the 100 and 200 series. The 300 series engines were also more powerful and offered up to 25 percent better fuel economy than the earlier versions. They were also much quieter and conformed to the latest internationally agreed-upon standard for airplane noise abatement.

Boeing built 1,114 of the earlier 100 and 200 series planes, the last of which rolled off the Renton, Washington, production lines in 1988. So-called "new generation" 737s like the 300 (and later) series were built in large volumes from

1985 on. Apart from the engines, the other significant change in the aircraft was an up-to-date cockpit, featuring more electronic instruments and, if the airline wanted them, computer screens instead of conventional clock faces, or "iron dials," as pilots of more modern aircraft dismissively referred to them. The 737-300 is also more than 9 feet longer than the earlier models and seats up to 19 more passengers. But in almost every other respect, it is the same aircraft as the 200 series.

USAir maintenance personnel checked out N513AU and passed it for flight after routine servicing. It left Bradley International Airport at Windsor Locks, between Hartford, Connecticut, and Springfield, Massachusetts, at 6:20 A.M. Its destination was Jacksonville, via Syracuse, Rochester, and Charlotte. Bruce Peck, who flew as co-pilot as far as Charlotte, told Chuck Leonard there had been no problems whatsoever with the aircraft.

Airlines like USAir work their 737s hard. With a 1994 sticker price of $30 million, the cost to an airline can easily double when financing costs are added over a ten-year period. Depending, of course, on how the airplane was financed, that means the interest bill alone is several hundred dollars an hour, possibly $5,000 a day, so every minute an aircraft spends on the ground, or in the hangar, represents a serious cash drain on the company. Airline schedules are designed as much to keep aircraft in the air, as they are to suit the convenience of passengers. Operating on the hub-and-spoke principle, smaller aircraft like Boeing 737s fly passengers from smaller cities into larger ones like Philadelphia where, if they have an onward, longer journey to the West cost or to Europe, for example, the passengers transfer to larger aircraft such as the 747, 767, or DC-10. Those bigger airplanes are uneconomical for shorter hops involving multiple takeoffs and landings, an employment for which the 737 is ideally suited. However, a passenger making repeated short 50-minute journeys along a spoke into a hub like Pittsburgh will rarely board the same 737 aircraft twice. Having the same airplane shuttling back and forth on a route could be uneconomic because the business on the route might only justify, say, two or three return trips per day, each occupying two hours, when turnaround times are taken into account. An airplane could be sitting on the apron for two hours in between flights, at a cost to the airline of hundreds of dollars in finance charges alone, not to mention the cost of having pilots and cabin crew twiddling their thumbs waiting for takeoff. Instead the schedule is designed to get that airplane and its crew off the ground to a different city and back into revenue service as quickly as possible. Schedulers cue flight departures according to when there will be an airplane and its crew finishing another one. Which is why Captain Germano and First Officer Emmett were criss-crossing the NorthEast and going as far south as North Carolina, rarely returning to the same airport twice in one day.

Chuck Leonard's Operations Group also traced the pilots who took the aircraft from Charlotte to Jacksonville, where it was handed over to Germano and Emmett. They told a similar story of a well-behaved aircraft and an uneventful flight. But a passenger on the last leg told a different story, recalling for one of Leonard's team an "abrupt maneuver" on the approach to Jacksonville. Leonard asked that the digital flight data recorder be re-examined about the approach to Jacksonville. It showed a roll of 9° to the left, followed by a bank of 12° to the right. The pilots, Jeff Overton and Randy Jones, said this was nothing unusual, adding that a slight roll sometimes occurred as they switched control of the aircraft from manual to autopilot.

Germano and Emmett took N513AU out of Jacksonville at 10 minutes past 1:00 P.M., bound for Charlotte. Also aboard was an off-duty USAir pilot bound for Chicago, Captain Bill Jackson, who sat in a passenger seat until Charlotte. The plane, now called Flight 1181, was fully booked out of Charlotte, so Jackson moved to the cockpit jump seat. They left Charlotte at 3:21 P.M. and reached Chicago's O'Hare International Airport shortly after 5 P.M. Andrew McKenna, a passenger homebound for Chicago and seated in seat 1C, heard a noise coming from the ceiling of the cabin. He later told the NTSB that it sounded like "water gurgling" down the drain of a sink. He mentioned it to a flight attendant who listened and said it was coming from the cabin PA system.

Captain Jackson recalled that Germano flew the aircraft into Chicago while Emmett handled radio traffic and called out the checklists. He found both men in good spirits; they appeared well rested and not tired or stressed. Their flying was professional, Jackson recalled, and there were no problems evident with the airplane. At one point a flight attendant contacted the captain and said there were PA noises getting into the cabin. Jackson looked down and noticed he had kicked the PA microphone stowed near his foot and inadvertently switched it on. He reset the switch and the noise apparently ceased.

Gerald Fox, a USAir maintenance foreman, returned from a meeting in Pittsburgh and checked in at the USAir maintenance office in Chicago before leaving for home. The phone was ringing, and as he was momentarily the only one in the office, he answered. A woman, who did not give her name, said her husband was booked to fly on USAir 427. After he checked in, she overheard people who were getting off Flight 1181 discussing unusual noises they heard during the flight, and she wanted to know if the noises would be looked into. Fox said there were two mechanics looking after the aircraft, and if there were any problems, they would be fixed before the plane took off. However, the mechanics on duty did not appear in the office, so he walked out to the jetway and met Captain Germano, who told him there were no maintenance problems with the airplane. Reassured, Fox signed out and went home for the evening.

Mark Kobut was the lead mechanic on duty in Chicago that afternoon, and he supervised the pushback of Flight 427 from the gate by a specially equipped tractor. He was the last person at Chicago, apart from air traffic controllers, to speak with Captain Germano, doing so via an intercom headset plugged into the nose of the aircraft. Everything, he said afterward, was totally in order. Another USAir mechanic, Timothy Molloy, performed a walk-around check of the aircraft, making sure that all the doors were closed, the tires were properly inflated and had no leaks, all the navigation lights worked, and so on. He also checked the pitot tubes and probes—sensors extending clear of the aircraft skin that are used to measure airspeed and outside air pressure and provide other flight information to the pilots. He found no blockages or other problems.

Aboard were 132 people, including the pilots and three flight attendants: Stanley Canty, April Slater, and Sarah Slocum-Hamley. Bob Van Auken, the USAir dispatcher who was responsible for ensuring that the flight was properly loaded and fueled, handed Captain Germano a sheaf of 22 computer-generated pages and a form to sign that detailed the number of passengers, amount of luggage, volume of freight, and quantity of fuel. Flight 427 would carry much more than enough fuel to fly to Pittsburgh, to deal with any holding delay or a diversion to another airfield. The papers contained detailed briefings on all the airports and weather conditions in his sector. Germano signed the form and the flight proceeded.

USAir Flight 427 taxied out to runway 32L at 1802 hours, and it was airborne at 1810. The flight was scheduled to take 55 minutes at a peak altitude of 33,000 feet, but as the aircraft climbed, air traffic controllers instructed the crew to go only to 29,000 feet because of conflicting traffic on the route. (Air traffic controllers attempt to keep aircraft at least 5 miles apart horizontally and at least 1,000 feet, sometimes 2,000 feet, apart vertically. Since 427 was a short flight, they decided to direct it to the lower altitude.) From the standpoint of the controllers and crew, the takeoff was just what they liked—uneventful and routine.

For his understanding of the remainder of the flight, Leonard depended on the Cockpit Voice Recorder Group and the Flight Data Recorder Group for help. The CVR work revealed that Emmett was flying the aircraft while Germano handled the radio. Up to the moment the aircraft departed from its assigned altitude, the FDR indicated both pilots had faultlessly navigated the aircraft and meticulously obeyed air traffic control instructions.

At 1843 hours, Germano asked Emmett if he wanted to "let 'em up for a while?" and this was followed by the sound of the seat belt chime sounding. Two minutes later Emmett is heard swearing at something in the cockpit. "Ah, you piece of shit," he said. At 1847 hours, Germano asked Cleveland Center, which was handling air traffic control for his sector, for permission to descend to 24,000 feet. He was given an additional altitude, 10,000 feet, at which to pass over a navigation beacon later in the flight.

At 1851, a broadcast from the Automatic Traffic Information System (ATIS) came in. This provided the pilots with recorded information on weather, visibility, altimeter settings, and the status of the navigation and instrument landing systems at the Pittsburgh airport.

At 1853 hours, a flight attendant came into the cockpit and asked if the tower had assigned a docking gate. Germano said, "Not yet". Then she said, "Do you know what I'm thinking about? Pretzels." She asked if they would like to try her special "fruity juice cocktail." Captain Germano wanted to know how fruity it was, and Emmett said he was glad to be a guinea pig.

Thirty seconds later, the Cleveland controller repeated the altitude information and told them to maintain a speed of 250 knots. Germano confirmed the controller's request. Emmett continued to have problems entering data into the flight management computer that feeds commands to the autopilot and the autothrottles. At 1856, Cleveland asked them to reduce speed to 210 knots, but Emmett thought Cleveland said 250 knots, and Germano said he might have misunderstood the controller. Shortly before 1857, Germano contacted Pittsburgh's controller and told him they were descending to 10,000 feet.

At 1857 hours, the flight attendant returned with the drinks. Both men tasted them and approved. Emmett said his would be good with some dark rum in it. Then the Pittsburgh approach controller cut in, telling them to take a heading of 160° to runway 28 Right and maintain a speed of 210 knots. Germano replied, "We're coming back to two-ten and one-sixty heading down to ten. USAir four twenty-seven and, uh, we have Yankee." Yankee was a reference to the earlier Pittsburgh ATIS broadcast.

The two men spent the next 40 seconds guessing the ingredients of the fruit drink, until an automatic cockpit alarm alerted them that they had now descended to 10,000 feet. USAir, like most US airlines, operates a "sterile" cockpit below 10,000 feet in which visitors to the flight deck are discouraged. "OK, back to work," said the flight attendant and left the cockpit. Then the Pittsburgh controller told them to descend to 6,000 feet. Just after 1859, Emmett suggested that they go over a preliminary checklist, and they reviewed their instrument settings for the last time. Then Germano asked for shoulder harnesses and there is the sound of a click like a seat belt being fastened. At 6 seconds to 7 P.M., Germano says, "Ah, don't do this to me." Emmett chuckles and asks, "Froze up, did it?" After 7:00 P.M., an intermittent static sound is heard on the Captain's channel for about 17 seconds. The Pittsburgh approach controller asks them to turn left 140° and reduce their speed to 190 knots, and Germano replies in the affirmative.

Then follows the sound of three clicks similar to the sound of the flap handle being shifted to extend flaps for the final approach. This is followed by the sound of the seat belt chime. Emmett then said, "Oops, I didn't kiss them 'bye," meaning he hadn't given the passengers the usual final spiel before landing. He asked Germano for the temperature given on the Pittsburgh Yankee, and was told it was 75°. In the background is heard the sound of the elevator trim wheel being turned by the autopilot. The flight attendant is heard over the PA system asking passengers to keep their seat belts fastened and to remain seated after landing, followed immediately by Emmett saying, "Folks, from the flight deck, we should be on the ground in about ten more minutes. Uh, sunny skies, little hazy. Temperature's, ah, seventy-five degrees. Wind's out of the west around ten miles per hour. Certainly appreciate you choosing USAir for your travel needs this evening, hope you've enjoyed the flight. Hope you come back and travel with us again. At this time we'd like to ask our flight attendants please prepare the cabin for arrival. Ask you to check the security of your seat belts. Thank you."

At 1 minute past 7:00, Germano contacted Pittsburgh air traffic control and asked if they had been assigned runway 28 Left. The tower replied, saying it was 28 Right. Emmett said, "Right, two eight right, that's what we planned on. Autobrakes on one for it."

There followed some more small talk about various instrument readings. On the radio they could hear other aircraft being instructed to make turns, and Germano said, "Boy, they always slow up so bad here." Emmett replied, "The sun is just like it was taking off in Cleveland yesterday, too. I'm just gonna close my eyes. You holler when it looks like we're close." And he laughed. Germano chuckled for the last time.

The tower told them to turn left and adopt a heading of 100°. The controller added that there was nearby traffic, a northbound Jetstream commuter aircraft 6 miles away, climbing to 5,000 feet. "We're looking for the traffic, turning to one zero zero," replied Germano.

Next was heard the sound of the engine revolutions steadily increasing and the clicking sound of the stabilizer trim wheel turning at a rate commensurate with autopilot operation. At 2 minutes and 54 seconds past 7:00, Emmett says jokingly in a French accent, "Oh ya, I see the Jetstream." He cannot know it, but he has just 28 seconds to live, 28 of the most mysterious and contentious seconds in aviation history.

At 2 minutes and 57 seconds past 7:00, three thumps are heard in the cockpit. "Sheez!" exclaims Germano. "Zuh," says Emmett almost simultaneously. Then there is another mysterious thump. Germano inhales and exhales rapidly. There's an unknown clicking sound, then another thump, this time softer. "Whoa," says Germano, and there's more mysterious clicking and Emmett grunts. There's another, different clicking sound and Germano yells, "Hang on!" The aircraft's engines power up and there's more grunting, louder this time, from Emmett. "Hang on," repeats Germano, and there's another click and the sound of the autopilot-disconnect warning, a wailing horn. "Hang on," repeats Germano. "Oh shit!" says Emmett. "Hang on!" repeats Germano again. Then there is a sound, increasing in intensity and similar to the onset of the buffeting that accompanies a stall. Then there's a vibrating sound, just like the sound from the stick-shaker, a device that causes the control column, or "stick," to rattle as a warning when the aircraft is in danger of stalling—where the wings lose lift and the aircraft falls from the sky. This sound continues until the end.

"What the hell is this?" asks Germano as various automated warnings go off in the cockpit. "What the ...," he repeats. "Oh," says Emmett. "Oh God, Oh God," says Germano. Then the approach controller cuts in and he's only just said "USAir ..." when Germano yells into the radio mike, "Four twenty-seven emergency!"

"Shit," says Emmett. "Pull!" yells Germano. "Oh shit," says Emmett, "Pull, pull," says Germano. "God," says Emmett, and Germano screams. "No!" yells Emmett and the recording suddenly ends.

Because there was an open microphone to the control tower in the final seconds, the terror and panic of the two pilots as death rushed toward them was also recorded on the air traffic control tape. Unlike the cockpit voice recorder tapes, of which only a transcript is officially released, the FAA routinely releases the recordings of air traffic controllers' radio traffic. On this occasion, in deference to the pilots' families, a decision was made to withhold the tape. "You don't want kids to hear their daddy die on the radio," said Alan Lebo, the former air traffic controller now chairing the NTSB's Air Traffic Control Group on the Flight 427 investigation.

Lebo can usually tell within two days of the start of an investigation whether or not any errors or shortcomings by air traffic controllers contributed to a crash. All controllers at major airports are employed by the FAA, which routinely records all radio traffic. The tapes are held for 30 days, after which they are recorded over. In the event of an accident, they are handed over to investigators along with printouts of all radar contacts with the crash plane.

In Los Angeles, on February 1, 1991, FAA tapes clearly showed how an overworked and poorly supervised controller had instructed a Metroliner commuter aircraft with 12 people aboard to taxi to the end of the runway and wait there for takeoff. The same controller then gave permission to an incoming USAir jetliner to land on the same runway. It landed atop the Metroliner, killing all aboard, then crashed into a building and burst into flames, killing an additional 21 people.

But in Pittsburgh on the evening of September 8, 1994, it was immediately clear that the controller on duty when Flight 427 crashed had made all the right moves. He was Richard Fuga, a man with 12 years' experience. According to his manager, he was "as good a controller as you're going to find, one of the top three or four in the building." Fatigue was ruled out when it was found that Fuga was on only the second day of his workweek. Prior to that he had two straight days off duty. He had been working on the Feeder Radar North since 6:15 P.M., guiding incoming traffic until it was close enough to be handed off to other controllers who would guide it to the runway. As often happens, there had been a "push" of traffic toward the end of the day, and Flight 427 was the last arrival from the northwest in that particular bunching.

Fuga instructed 427 to turn to a 100° heading and to maintain 6,000 feet altitude. However, when he next looked at the cluster of numbers that represented the information he had on the flight, it had descended to 5,300 feet, and he radioed to tell it to remain at 6,000 feet. But as he did so, Germano was commencing his emergency broadcast. Fuga heard some unintelligible words, then some swearing. He made his altitude request, then heard Germano yell out "four twenty-seven emergency," plus some unintelligible comments. Fuga waited a few seconds, then called, "USAir four twenty-seven Pittsburgh." He repeated the call several times without response, then noticed to his horror that the radar blip and the numbers that had been Flight 427's "squawk" code and altitude were now replaced with a series of Xs on his radarscope. He called again, "USAir four twenty-seven radar contact lost," but there was no reply. Fuga alerted his immediate supervisor and handed over his traffic to the arrival controllers.

Kenneth Erb, Fuga's area supervisor that night, was returning from a break when he was told there was "a bad emergency, a target has disappeared." Erb, a veteran of 32 years in air traffic control, halted all departures and arrivals and put out the alert on all frequencies. Five minutes later, the tower called back and said a plane had crashed in Hopewell Township. After calling the airport emergency services, Erb allowed arrivals to resume but kept departures halted for awhile longer to give elbowroom to emergency vehicles rushing off the airfield. Erb then relieved Fuga from duty. "Must have been a bomb," Erb thought as he tried to piece together the frantic events. Otherwise the pilots would have been able to say what was wrong. Nothing else could happen so suddenly.

Besides a bomb, Al Lebo and the NTSB had other possibilities to chase down. One was a collision with birds. Another was a collision with a light aircraft, a glider, or even a balloon. Pittsburgh had a relatively new radar system, the ASR-9 (for Airport Surveillance Radar). Unlike older systems that reflected almost everything in their paths and often gave confusing readings known as clutter, modern Secondary Surveillance Radar systems like ASR-9 are more discerning. Lebo was told that the ASR-9 Fuga was using that evening was specifically designed to filter out stationary and non-aircraft targets, although one technical expert told him it would portray a flock of birds if it was large enough and moving fast enough. Others were not so certain. However, he was assured that nonmetallic moving objects, like gliders and balloons, would be displayed on the scope.

Birds, generally the lightest of vertebrates, are a serious aviation hazard. Worldwide, more than 3,000 collisions of birds and aircraft are reported each year, although the FAA suggests that perhaps 80 percent of bird strikes go unreported. The vast majority of incidents are minor, but occasionally these collisions wreck engines and other vital components. According to one estimate, bird strikes have destroyed more US Air Force aircraft in the past 25 years than have enemy actions. Air Force records state that there have been 38,000 collisions between birds and aircraft since 1985, in which birds destroyed 30 aircraft, killing 33 crew members and causing more than $500 million in damage. Worldwide, bird damage costs some $400 million annually, and a total of at least 400 deaths have been attributed to bird strikes. According to one estimate, a large adult gull striking an aircraft moving at 130 knots has an impact equivalent to 2 tons. Double the speed of the aircraft and you quadruple the impact. Even a flock of starlings is capable of doing immense damage to an aircraft.

In fact, starlings are "feathered bullets" with a body density 27 percent higher than herring gulls, according to the Bird Strike Committee USA. In October 1960, a Lockheed Electra crashed after colliding with a flock of starlings at Boston. Of the 72 people aboard, 62 died. In September 1987, a US Air Force B-1 bomber on a high-speed, low-flying training mission crashed after a large bird struck the wing near where it joins the fuselage and the aircraft's hydraulics systems were incapacitated. Only three of the six crew aboard bailed out successfully. In other accidents, large birds have penetrated the cockpit after first crashing through the radar installation in the nose.

Lebo had to consider the possibility that a bird, or several, could have smashed through 427's cockpit window, killing or incapacitating one or both of the pilots, or been ingested by the engines, knocking one or both of them out of action.

He carefully scrutinized the radar printouts supplied by the FAA from the Pittsburgh control center for any evidence of an unidentified object. He found nothing, but that in itself was not conclusive, given the differences of opinion on the radar's ability to detect birds. Large, fast-moving flocks could perhaps be spotted, but one, two, or three individual birds? Nobody was sure. Ruling out collisions with gliders, ultralight aircraft, and balloons was easier: none were reported missing in the days following the crash.

Wake turbulence was another possibility Lebo had to consider. A smaller aircraft taking off or landing behind a DC-10 or a Boeing 747 will often get rolled about as it enters the slipstream of the larger plane. Wake vortices, rotating tubes of air coming off the first aircraft's wingtips, are what create the danger. Although it had long been a recognized hazard for light aircraft, wake turbulence had been discounted as a hazard for jetliners until May 30, 1972, when a Delta Airlines DC-9, similar in size to a Boeing 737, crashed onto the runway at Greater Southwest International Airport in Fort Worth, Texas. The DC-9, on a training mission with four pilots aboard, was following a larger American Airlines DC-10 that was doing landing exercises. When the DC-9 flew into the wake vortices coming off the left wing of the DC-10, the DC-9 flipped on its side, the left wingtip struck the runway, and the aircraft somersaulted onto its back, killing all aboard. The almost still weather, with just a gentle crosswind, contributed to the accident because it allowed the vortices to remain suspended above the runway instead of decaying or blowing away downwind out of harm's way.

Lebo searched the Pittsburgh departure and arrival records for evidence of larger aircraft notorious for having "dirty" wakes that could have been in Flight 427's vicinity at the time of the crash. There were no 747s or DC-10s, and the only other suspect was a departing British Airways Boeing 767. The closest it came to Flight 427 was 38 miles, too far to have an effect. A Delta Airlines Boeing 727 was flying into Pittsburgh 4.2 miles ahead of Flight 427. Slightly larger and with a similar wing profile, the 727 was not regarded as a prime suspect, although Lebo did note that Flight 427 had to pass through the same spot in the sky that the Delta jet had vacated only seconds earlier.

Among the many possibilities Chuck Leonard had to consider as Operations Group head was the likelihood that Flight 427 might simply have run out of fuel and fallen from the sky. It was a possibility he investigated meticulously despite the rather obvious evidence from the crash site of a conflagration that consumed much of the wreckage, and withered, discolored vegetation that was easily traced to large amounts of aviation fuel erupting from split tanks. There had been some near disasters involving low fuel levels. Four years earlier, USAir had eliminated some pre-departure fuel checks that management at the time considered excessive and unnecessary. In the 14 months prior to the crash, nine USAir flights were found to have left the gate with less than the amount of fuel required by company policy. One of them had to declare an emergency and land at an alternative airport because it was running dangerously low on fuel.

But a painstaking search of dockets at Chicago revealed that Flight 427 had, in fact, taken the correct amount of fuel aboard. So another possible cause of the crash could be eliminated. Another week passed inconclusively. Yet the list of possibilities seemed to be growing, not shrinking. Not long after the accident, Boeing pulled their only representative, a test pilot, out of the Operations Group and reassigned him to Albert Reitan's Cockpit Voice Recorder Group, where he became Boeing's second pilot representative. Chuck Leonard suspected that Boeing was wagering that he would not trace the cause of the crash, and his group would soon be wrapping up its work without reaching any conclusions on the causes of the crash of Flight 427. Somewhere among the unexplained exclamations, clicks, and thumps on the cockpit voice recorder tape might lie the vital clue that everyone else had missed. Nonetheless, it seemed to Leonard that Boeing was not so much interested in eliminating possible causes of the crash as in dashing to be where it thought the action was going to be. He wondered why.

Chapter 8 SITUATION: NON-NORMAL

On July 13, 1993, Boeing issued a 6-page document as part of a series of occasional newsletters called *Boeing Flight Operations Review*. The new edition had an unwieldy title: "Guidelines for Situations Which Are Beyond the Scope of the Non-normal Procedures." The average person would call a non-normal procedure on an airplane an emergency of some kind or other.

Evidently, Boeing intended its guidelines to cover things that were more dire than mere emergencies, like a midair collision, a bomb explosion, "or other major malfunction." As an example of a major malfunction, it listed jammed ailerons or rudders, and offered the following reassuring advice: "If a jammed flight controls condition exists, both pilots can apply force to either clear the jam or activate the break-out feature designed into all Boeing airplanes. There should be no concern about damaging the mechanism by applying too much force."

The Boeing document was neither a "must do" Aviation Directive issued at the behest of the FAA nor an urgent warning about some newly discovered fault. Instead, it was written in the soothing language that Boeing, like many corporations, uses to reinforce confidence in the company and its products, even if the message is about something dreadfully unpleasant.

After the 1994 crash of Flight 427, Chuck Leonard, the NTSB's Operations Group chief, decided to carefully compare the 1993 "Situations ..." article by Boeing with USAir's manual for flying the 737, which had not been updated since 1990. This ring-bound manual, provided to each 737 pilot, conveyed USAir's instructions on how it wanted the airplane flown, and the airline made clear that it wanted its instructions to be followed to the letter. The problem was, the manual did not deal with the sort of situation Captain Germano and First Officer Emmett were faced with in the sky over Aliquippa shortly before Flight 427 plunged to earth. What do you do when you suddenly find your aircraft on its side or upside down? How should a pilot react to a rudder that inexplicably goes into hardover? Is there any way to pull out of a nose-dive from extremely low, runway-approach, altitudes? About situations like these, which, to say the very least, fall within the category "non-normal," the USAir manual, which focused on normal operating procedures and well-understood emergency situations, was silent.

USAir was not alone in failing to address some of the issues raised after 1991 by United Flight 585; most US airlines were in a similar position. The exception was United Airlines, which in 1992 started to face head-on the question of what pilots should do when something happened that their training had not prepared them for. This new and unorthodox training initiative had already been under consideration when United Flight 585 crashed, but that catastrophic event gave new meaning and urgency to the notion that something more than flying "by the book" might be necessary.

"Inadvertent upsets have occurred in the airline industry with catastrophic results," ran United's introduction to a new 1994 pilot training manual called *Advanced Maneuvers Package*, in language much less circumspect than Boeing's. In certain respects, the airline was calling for the sort of "seat of the pants," almost intuitive, flying that previously had been discouraged. The major airlines' emphasis had been on flying in such a way as to avoid emergencies, rather than on what to do if and when one happened. Standard Operating Procedures were the order of the day, and there were official checklists for everything, including well-known emergencies. But the notion of hauling out a clipboard to follow a tightly prescribed procedure when an aircraft is barrel-rolling about the sky is absurd. By August 1994, a month before Flight 427 crashed, United had completed the software for its *Advanced Maneuvers Package* and was phasing it into the normal simulator training for the larger Boeing 757 and 767 airliners, with one or two maneuvers introduced at the end of each training session. New pilots and pilots new to the aircraft received the training first, and it was given to existing pilots at the routine refresher training sessions.

Intrigued by this innovation in pilot education, Chuck Leonard and his boss, Tom Hauteur, flew to United's training facility in Denver in late November 1994, after the crash of 427. They met with Captain Larry Walters, United's man in charge of the advanced maneuvers program, then took a two-hour simulator session to try out some of the maneuvers. Walters believed that the traditional tight rein kept by airlines on their pilots, limiting them to a very narrow band of permitted operations, had limited their confidence in their ability to react to events "outside the envelope."

The new maneuvers demonstrated the extreme caution with which traditionally trained pilots approached emergencies. In a stall, for example, the nose of the aircraft must be pushed down rapidly to increase speed, to boost the airflow past the wings, and to build up lift. But pilots were traditionally trained not to make dramatic maneuvers and were not fast enough in pushing the nose sufficiently far below the horizon. Rather than having their knuckles rapped for getting into a stall during simulator training, United pilots were now being encouraged to bring their "aircraft" to the point of stalling so that they could see the ease with which a full stall could be avoided. Another maneuver showed pilots how to escape from a practically inverted condition, using a combination of aileron and rudder. They were also shown methods for getting a plane back to level flight after enduring combinations of extreme bank angles and precipitous nose-up and nose-down situations. Finally, they learned what to do in case of engine failure when it was most serious, like after aborting a landing. All of these situations had one thing in common: the need to react quickly, and without reference to checklists of standard operating procedures.

Far from slowing up simulator training, Leonard was surprised to learn that United's new advanced maneuvers training was improving overall student performance. The pilots were doing better not just on the "intuitive" tasks, but on the checklist procedures as well. This was an unexpected payback from the program, which was to be extended to all aircraft types, including the 737, as soon as instructors could be trained. On the way back to Washington after the visit to the training center, Leonard wondered aloud to Hauteur that if Germano and Emmett had been similarly trained, would Flight 427 have landed safely at Pittsburgh? Of course, Hauteur could not answer the question—he didn't even know, at this point, what caused 427 to spin out of control in the first place.

As his investigators continued to draw blank after blank in their search for the cause of the crash, Hauteur decided to order Cynthia Keegan to mount a reconstruction of the wreckage of Flight 427 in the hope that it might offer clues, and possibly a solution. And, realizing the enormity of the task ahead of her, he asked an old friend, Dave King, to lend a hand. King was the principal inspector of air accidents at the Air Accidents Investigation Branch (AAIB) in the UK. He was a veteran English crash investigator who had probed the UK's worst aircraft disasters, including the 1988 bombing of TWA Flight 103, a Boeing 747, over Lockerbie, Scotland.

The AAIB and the NTSB routinely cooperate, often sending observers to each other's investigations. Sometimes this is mandatory: international protocol dictates that whenever an aircraft from one country crashes in another, an investigator from the first country is invited to join the investigation. But the cooperation in this case was voluntary because the crash of USAir 427 was a purely US affair. King offered some unique talents to the NTSB Structures Group, which had not recently attempted a reconstruction of anything so large or so badly smashed up. What's more, by inviting King to join the reconstruction effort, Hauteur was drawing on a 40-year-old tradition, for it was the English who had pioneered reconstruction as a crash-investigation tool.

On January 10, 1954, a British Overseas Airways Corporation Comet broke up in midair over the Mediterranean shortly after taking off from Rome. All aboard were killed. It was the third fatal Comet accident, and the second midair breakup, in just two years. The accidents cast serious doubts over the safety of the world's first jet-powered airliner and led BOAC to temporarily ground the remainder of its Comet fleet pending investigation. Three months later, just two weeks after the UK government had given BOAC the all-clear to resume flying Comets, there was another midair breakup about 350 miles further down the Italian coast from the January crash site, and the UK government fully withdrew the Comet's operating certificate. Wreckage from the January accident was dredged from the seabed over a period of several months and painstakingly attached to a full-scale mock-up. At the same time, a spare aircraft was placed in a huge, specially constructed water tank and repeatedly pressurized and depressurized to simulate repeated takeoffs and landings. When the test aircraft's fuselage eventually split along a window line, the investigators turned to the reconstruction and discovered that the crash plane had split at a similar spot. Case solved, with the British faith in reconstruction as an investigative tool firmly in place.

Dave King was shocked when he walked into the A1 hangar at Pittsburgh International Airport, where the wreckage from USAir 427 was being sorted. Investigation personnel, wearing biohazard suits, were sifting through airplane wreckage—some of it in a row of dumpsters, some of it lying on the floor. Nothing in his experience of investigating civilian air crashes had prepared him for what he now saw. Instead of large chunks of easily recognizable wreckage—perhaps complete sections of fuselage or large areas of wing or even sizeable areas of flooring—he was faced with tin kicking at its most difficult. Apart from a section of tailplane, a few sizeable chunks of wing spar, and some badly damaged landing gear, there was very little he could immediately assign to any particular part of the airplane. In fact, there was little in the hangar larger than a dinner plate. The last time he had seen wreckage this badly smashed up had been at the site of a high-speed jet fighter crash. He just could not believe that a large passenger jetliner could be shattered into so many small fragments.

While King was in Pittsburgh, there were four main phases under way in Flight 427's reconstruction:

1. A reconstruction of the forward pressure bulkhead, the front "wall" of the cockpit below the windshield and just in front of the pilots' legs. In front of that is the radar installation, located in the conical nose, or radome, of the aircraft. This reconstruction was designed to show whether or not a bird had penetrated the nose of the aircraft and entered the cockpit, disabling one of the pilots or jamming his flight controls and precipitating the crash. About 40 percent of the bulkhead was recovered and subjected to ultraviolet light inspection for blood or other traces of birds. Nothing was found there, or on parts of the wings where bird strikes might also be expected.

2. A reconstruction of the beams that were the main structural support of the airliner's floors, to see if these could have failed in any way prior to the crash. When an improperly latched door to the cargo hold of a Turkish Airlines DC-10 fell off shortly after taking off from Paris on March 3, 1974, it led to the rapid depressurization of the cargo hold and the collapse of part of the floor, severing all hydraulic lines and control cables to the tail. The pilots lost an engine and were unable to use their elevators to gain height, and the aircraft crashed in a forest, killing all 346 aboard. It was a carbon copy of an incident aboard an American Airlines DC-10 21 months earlier when the same door fell off in flight, though the pilots skillfully managed to land their crippled airliner in Detroit, and there were no fatalities. Very little could be satisfactorily deduced from Flight 427's floor beams. Forward of the wings, sometimes less than 5 percent of a beam could be identified. Things improved toward the tail, where upward of 95 percent of some beams were recovered. However, in some cases investigators were not sure if they could identify a specific beam, so alike and so badly damaged were many of them. They were luckier with the doors. Enough locking mechanisms and door frame wreckage turned up to enable them to deduce that all doors had been properly closed in flight.

3. An investigation of the empty auxiliary long-range fuel tank. This fuel tank was in the cargo hold and offered two deadly possibilities. One, that the tank had somehow collapsed due to a fuel pump being inadvertently switched on and left running, creating a dangerous vacuum in the tank. Two, that there were fuel vapors in the tank despite the fact that the tank was not in use, and that it had somehow exploded.

Either way, there was a possibility that the floor beams above it could have been affected, leading to the severing of control cables and hydraulic lines to the tail. However, King noted from drawings supplied by Boeing that the tank was not bolted to the floor beams, and he believed the possibility of the floor rupturing in that way was very remote.

4. A reconstruction of the wheel-well area. Brakes that were overheated could cause a fire. And the disintegration of rapidly spinning wheel rims as the undercarriage was retracted into the wheel well could cause structural damage and the severing of hydraulic lines and control cables. Investigators were particularly looking for a splattering of melted rubber in the wheel well, which might indicate a tire overheated or on fire as it was retracted. Some melted rubber was found, but it was insignificant and could have resulted from any number of causes, including a tire bursting on impact or in the fire.

King was pessimistic that anything positive could emerge from the reconstruction. This was not a reconstruction in the Lockerbie sense, where a three-dimensional mock-up of the fuselage was produced and pieces of wreckage pinned to it in the appropriate places. Flight 427's reconstruction was two dimensional. Full-scale drawings of plane body parts were spread out on the floor of the hangar and covered with transparent Plexiglas to protect them, and pieces of wreckage were laid over their appropriate places on the drawings. A first sift of the material in the dumpsters "resulted in a disappointing lack of positively identifiable components," as King noted upon observing how the investigators ran out of material to place on the drawings after a week. Another sift followed but it produced relatively little of value. "You might have a piece of wing skin but you would be hard-pressed to tell what part of a wing it came from, let alone which wing," King recalled.

King had to leave Pittsburgh in the midst of the reconstruction, and his report to Cynthia Keegan reflected his awe at the enormity and difficulty of the task she faced. "As with the floor beams," he wrote, " a further sift of the debris in the dumpsters will almost certainly fail to provide enough to furnish conclusive answers to the questions which the reconstruction is intended to address." But Keegan, who would never forget her helpless feeling when she first viewed the wreckage site, now believed some sense was being made of the tangled pile of shattered aluminum and metal alloy in the dumpsters, even if less than 10 percent of some parts and no trace of others could be located. For example, some of the cables leading to the flight controls were identified, but Keegan was left with four bins filled with unidentified cable fragments. Only 60 percent of the right rudder cable was found, and just 22 percent of the left one. Moreover, all the breaks in the cables were found to have been caused by the impact, and not by an in-flight failure that might have precipitated a crash.

Keegan was not discouraged that the reconstruction was incomplete. She knew what she needed to find, and believed what they'd already identified was more than enough:

> ... There was an assumption that a piece of the airplane might have fallen off prior to the accident, and we wanted to make sure that we had all the important parts, especially if there was any likelihood that they may have contributed to the accident.... I think we got all the important pieces. We were able to determine that all the doors were there, and from nose to tail and from wingtip to wingtip, we were able to determine that all the more important parts were. It didn't matter that we weren't able to identify all the skin panels—you'd have been pretty hard-pressed to say if something like that contributed to the accident in any event....

Overall, roughly half of the airplane was identified and laid on the floor. One by one, her team ran down the various possible causes of a crash and eliminated most of them, as much as they could, given the appalling state of the wreckage.

A typical investigation group during a plane crash probe is chaired by an NTSB specialist aided by one or two colleagues, and perhaps by representatives of the parties to the investigation—the airline, the plane maker, the pilots' and machinists' unions, and so on. In the case of 427, however, there was an exception for the first four months. The FBI had the run of the investigative activities, and ran its own investigations, always without the supervision and sometimes without even the counsel of the NTSB.

Any crash involving an aircraft that crosses state lines is likely to attract the attention of the FBI, especially if the crash cannot be immediately explained. And the fact that Flight 427 had on board a government drug informant by the name of Paul Olsen only made the disaster all the more interesting to the FBI. Special Agent William Perry, head of the FBI's Pittsburgh Bureau, arrived at the crash scene on the evening of September 8 and, with his colleagues, maintained a watch over the investigation. Later in the month he twice arranged the handover of pieces of wreckage and items of luggage to the FBI's laboratory in Washington for tests to see if they contained residues of explosives. The tests proved negative.

The FBI was not the only one with an interest in explosives. Members of the FAA's Aviation Explosives Security Program are immediately dispatched whenever word reaches it of an air crash for which there is no obvious reason. A member of the unit, Edward Kittel, arrived at the crash scene in the early hours of the morning of September 9 and was joined a few days later by one of his colleagues, Calvin Walbert. Kittel's first priority was to examine the crash site for evidence of explosives before the wreckage was removed because new sanitation regulations required wreckage be washed in a bleach solution, a process that could destroy vital evidence.

Kittel, a former military bomb disposal officer, had also received training in nuclear and biological warfare decontamination, and proved useful in other respects. He helped design and set up the decontamination stations through which all personnel and equipment had to pass when leaving the crash site. He then attached himself to Keegan's Structures Group and insisted on examining each significant piece of wreckage before it was removed from where it lay. Later he positioned himself at the staging area where wreckage was being collected before decontamination and checked it again before the bleach solution was applied. He was looking particularly for puncture holes and pitting on metallic surfaces that might indicate a bomb blast, or something entering the aircraft from the outside, like a missile or a bullet. His colleague, Calvin Walbert, focused on luggage and personal effects.

They made an unusual request of the hard-pressed coroner, Wayne Tatalovich, asking him to x-ray all the human remains that passed through his morgue and remove from these every piece of foreign matter that showed up on the x-ray plates. The thinking was that if there had been an explosion, any bodies close to it might have "captured" bomb parts or fragments. Tatalovich complied, and the result was buckets of fragments, which Walbert dutifully sifted through without finding anything suspicious.

After the wreckage was moved to the Pittsburgh hangar, Kittel and Walbert renewed their forensic probe. This time they looked at the edges of the shattered metal fragments as the giant, incomplete jigsaw took shape on the hangar floor, and tried to answer the question: what broke it up in the first place? Two adjoining pieces of wreckage with the mating edges splayed outward (or inward) as though something had blasted through it would have been highly suspicious, but nothing of the sort showed up. All the breaks they examined could be explained either by impact with the ground or from hitting trees on the way down.

Kittel and Walbert went back to Washington, but Kittel was called back to the Pittsburgh hangar on November 17 after the remains of the floor beams and the control cables that ran beneath them had been assembled. Cynthia Keegan wanted to know whether a bomb could have caused the damage to some of the beams and any of the several breaches visible in the cables. Kittel marveled at the rate at which the reconstruction had progressed since his last visit, but saw no evidence of an explosion.

The bombing of an aircraft of course attracts massive publicity, which is largely why terrorist groups do it. In the three months after the crash, no organization had claimed responsibility for the downing of Flight 427. But, in the absence of any other cause emerging, the remote possibility of a bomb nagged Tom Hauteur. In the New Year he would have to give a report at an official public hearing on Flight 427, and while it now looked as if he might not be able to say why it crashed, he at least wanted to be able to list the areas of investigation that were still active. Could he at least cross a bombing off his list of possibilities? The FAA's Kittel had concluded that bombing could be ruled out, and Keegan relayed the details of Kittel's reasoning to her boss. But Hauteur wanted a second opinion, and this time he turned to the FBI.

On December 19, 1994, Supervisory Special Agent Wallace Higgins from the FBI's Washington explosives unit met in Pittsburgh with a team of three FBI explosives experts drawn from all over the country. Over the next two days the four men pored over the wreckage, examining thousands of pieces of debris. They found no evidence of an explosion. For almost the first time in the investigation, Hauteur had something definite to report. He could not say what caused the crash, but he could put his hand on his heart and declare that Flight 427 had not been brought down by a bomb or a bird.

ꞏapter 9 PUMPS AND VALVES

Captain John Cox was sitting in a Pittsburgh restaurant with a colleague from his union, the Air Line Pilots Association (ALPA), when they learned that USAir Flight 427 had crashed. Cox was a USAir pilot who flew the 737, and he was an active member of his union's aviation safety committee. He had been in town for an ALPA meeting with USAir. His dinner companion, also in Pittsburgh on union business, was Captain Bill Sorbie, chairman of his air safety committee. As well as dealing with ALPA flight safety issues, Sorbie also allocated pilots to take part in air crash investigations under the NTSB party system. Now he had a major accident right under his nose. Both men were stunned and incredulous. This was USAir's fifth major accident in five years, and its second in 1994.

The wreckage, less than two hours old, was still burning when Cox and Sorbie were allowed through the yellow police tape and past the cordon that had been thrown around the site. The scene was apocalyptic. The darkness was seared by the bright blue-white light of arc lamps atop tripods, supplementing the cab-top spotlights of the fire trucks and other emergency vehicles. Flashing blue lights strobed everywhere. These were punctuated by fitful bursts of yellow flame as the fire continued to defy the best efforts of the firefighters.

Back at his hotel, Cox called his wife at their home in St. Petersburg, Florida, and asked her to send him some things he needed. She was a flight attendant and was familiar with the routine—she'd been through the union's participation in crash investigations before. "See you in six months," she said, half seriously. Cox first became involved with investigations in 1986, when a pilot friend of his failed to stop his aircraft before the end of a runway and ran his plane into the mud at 60 knots. Nobody was seriously injured, but the multimillion-dollar aircraft was a write-off. Helping his friend assemble technical data for his defense, and becoming intrigued with the investigation process, Cox set about learning more. Before long he was known as a "safety guy" and was encouraged to take a more up-front role by ALPA.

USAir, like many other airlines, allowed special concessions to an agreed-upon list of pilots working on union business. Cox could phone ahead and get relieved of flying duty in order to attend meetings or help with investigations. He continued to be paid, although the union was required to reimburse the company for his time. Often he continued to work for the union at his own expense on days off or during pre-arranged leave. Flight 427 would be Cox's fifth major accident investigation for the union, and it would consume huge chunks of his spare time for the next five years.

Cox and Sorbie next made a few calls to see what else they could learn about the crash, in particular whether the pilots had reported anything amiss to air traffic control before the accident. But at this early stage, little was known. And as the hours went by, both men had the sense that this looked like an investigation that might have to go the full course, and this in turn meant that they and ALPA were in serious trouble, organizationally speaking. A major crash might demand the contributions of a dozen people from ALPA, and Cox knew for certain he did not have that many trained volunteer investigators. Sorbie asked Cox if he would help out by working on this latest crash, and Cox agreed. But along with some of ALPA's best people, Cox was still working on the case of a DC-9 that had crashed in North Carolina just two months earlier. For sure, he was going to be stretched thin.

USAir flight 1016 had come down in a thunderstorm after aborting a landing, killing 37 of the 57 people aboard and seriously injuring most of the 20 survivors. The crash was turning out to be a difficult investigation. Weather in the form of microburst-induced windshear was the major factor, but serious shortcomings in other areas were being uncovered. There were major questions about how the pilots responded to the emergency, the USAir procedures set in place for handling such events, air traffic control's management of the affair, and even the reliability of windshear detection equipment aboard the crashed aircraft. All of this preyed on Cox's mind as he tried to focus on what had just happened to 427. The next morning in Pittsburgh, Cox got up early to attend the first 427 organizational meeting, where Tom Hauteur assigned him as the ALPA representative to the Systems Group. Now he had two crashes competing for space in his head.

There was for Cox, though, at least one thing to be grateful for. The leader of the Systems Group was Greg Phillips, someone with whom Cox had worked before—and liked and respected. Compared to some of his colleagues, who looked as if they were under considerable pressure, the slow-speaking and extremely courteous Phillips appeared almost unduly relaxed, but Cox knew he was one of the NTSB's sharpest and most experienced investigators.

Phillips had been consumed by aviation since he was a schoolboy in Indiana. As a teenager he had spent every spare dollar he earned—at a part-time job in a local Burger King—on flying lessons. When he was still in high school, a nearby college basketball team was wiped out in a plane crash, and when a group from the NTSB arrived to probe the incident, he eagerly followed its progress in the local paper. A few years later, Phillips had a personal encounter with the agency: an NTSB investigator questioned him about a small airplane he had rented in the course of training for his pilot's license. The next person to rent the plane died when it crashed, and the NTSB man wanted to know if Phillips had noticed anything strange about the aircraft, especially the stall warning system. No, Phillips told him, the plane seemed fine, and that was that.

After earning a degree in mechanical engineering and management from the University of Evansville, in Indiana, Phillips went to work for Cessna, working on structures, systems, and controls. He also joined the company's flying club and earned his commercial certificate and instrument rating, as well as qualifying as a flight instructor. Next he went to Los Angeles to work for the Northrop Corporation, designing systems for the F-18 fighter and starting work on a master's degree in management. It was exciting being at the leading edge of airplane technology, but it was a classified project, subject to strict military secrecy. When he answered a help-wanted ad from the NTSB, which was looking for systems engineers, he realized, as he put it afterward, "I was getting tired of not being able to talk about my work." At first he wondered about the wisdom of becoming a civil servant, which he suspected would pay poorly, not be very satisfying, and put him at the whims of politicians. He was right only about the pay.

Phillips's first major investigation involved a United Airlines DC-10 that crashed and burned at Sioux City, Iowa, in July 1989. The accident, which killed 111 of the 296 aboard, was horrific and yet one of aviation's most remarkable survival stories. When the engine in the tail disintegrated in flight, it severed all the aircraft's hydraulic systems, including the standby system, leaving the pilots without any flight controls whatsoever. There were no ailerons, no spoilers, no flaps, no elevators, no rudder. There was no way of making the airplane turn, climb, or descend, no way of extending the wing area using flaps so that the plane could slow to a safer speed for a landing. However, the pilots ingeniously used the throttles on the two remaining engines to control the aircraft. By speeding up the left engine and slowing the right, the aircraft could be put into a right turn. A left turn was accomplished by reversing the maneuver. The aircraft could be forced to climb by increasing power on both surviving engines, which tended to point the aircraft's nose up because they were slung beneath the wings. Descending was accomplished by reducing engine power. As they came in for a landing at almost double the normal speed, an unexpected crosswind caused the aircraft to bank suddenly. Without ailerons or rudder to oppose the bank, a wingtip struck the runway, causing the giant airplane to cartwheel and land on its back. It broke into several pieces, killing many passengers in the impact. Others were killed by smoke inhalation and fire after escaping fuel ignited. But almost two-thirds of those aboard survived, including the pilots, who were able to help Phillips and the other investigators reconstruct the event.

What had happened to the DC-10 was this: A defective fan disc in the tail engine broke up in flight, and pieces of it sliced through not one but three closely packed hydraulic lines in the tail. All of the aircraft's flight controls—the elevators, rudder, ailerons, spoilers, and flaps—plus the landing gear could be operated by hydraulic power from any one of those lines, although two of them normally operated in tandem, with the third in reserve as a backup. Phillips's work on the crash highlighted the fact that the pilots had had no fall-back, despite the existence of the third, backup, hydraulic system. All three control systems lines were severed, and the hydraulic fluid leaked away. The aircraft could have been landed safely if the pilots had any hydraulic power remaining, but they did not. Subsequently, McDonnell Douglas ordered the fitting of special valves on hydraulic lines that would close if sensors detected the rapid leakage of fluid, thus preserving some emergency power to operate the flight controls.

In 1991 and 1992, Phillips worked on the crash of Flight 585 at Colorado Springs. The experience left him frustrated. There were arguments for peculiar weather, to be sure, but there was also the nagging possibility that something went wrong with the rudder system or other flight controls. Overall, there was nothing strong enough to establish a probable cause of the crash. Despite his frustration, he knew the worst thing an investigator can do is pick the wrong thing as a cause, especially if the same type of accident happened again. "I would rather work hard, get to the point where we know we haven't enough information, but keep an open mind to see if it comes again," he was fond of saying. "There's nearly as much success in saying we really don't know what caused it, and we want to look some more."

Even though a report had been completed on United 585 on December 8, 1992, and the investigators assigned to other accidents, Phillips knew that the NTSB had never officially and completely closed the investigation. Since the publication of Flight 585's inconclusive final report, he had quietly compiled a list of Boeing 737 loss-of-control incidents that defied explanation in the hope that some sort of pattern might emerge. But what he found in this list of incidents was difficult to work with because so little of the data was captured on the flight data recorders. (Most of it had simply been written down in reports by the flight crew.) Phillips also became convinced that many problems in his 737 incident list were overstated. "When you're in an airplane, anything unusual that happens is bad, but maybe not as bad as you think," he explained. "We found that when we did have data, we could divide by three. If a witness said a 45-degree roll, it would turn out to be more like 15." By and large, the reports he came across did not involve an actual loss of control of an airplane; rather they involved incidents that resulted in little more than apprehension for crews and passengers.

As he briefed John Cox and the other members of his team prior to visiting the Pittsburgh crash site, Phillips put out of his mind the possibility that there might be any connection between the incidents on his list and Flight 427. He decided from the outset that he would direct his group toward a methodical examination of all the evidence offered to them, just like cops at a crime scene. And they would start in the "hole."

At an organizational meeting in the command center, a local hotel meeting room, Tom Hauteur laid out the basic ground rules. This would be, Hauteur declared, an open investigation. The various groups would agree on strategy and share their data. No one person was going to dominate this investigation, there would be no solo operations, and all data would ultimately be published in a publicly accessible form. Although many members of the various groups might have been sent in by outside parties, they were to regard themselves as NTSB investigators first and foremost and leave their party allegiances behind.

At the crash site, Phillips's Systems Group worked closely with Cynthia Keegan's Structures Group in tracking down and identifying components that belonged to their area. John Cox concentrated on the cockpit—as much of it as he could find—trying to find out what position various controls and dials might have been in at the moment of impact. They found an airspeed indicator, the radio magnetic indicator, some hydraulic gauges, some switches, and some of the warning lights. Then they concentrated on the flight controls and the tangled cables and hydraulic components that powered their movement.

Phillips, Cox, and the other members of the Systems Group were in the "hole" for a full week. A typical day began at 6:30 A.M. with a 7:30 meeting at the command center, in the hotel, for breakfast. At 8:00, there was usually a meeting of the Systems Group chaired by Phillips, where the coming day's work would be planned. They were then bused to a temporary facility near the crash site where they dressed in biohazard suits and gloves. After suiting up, the investigators were taken on a half-track rough-terrain vehicle into the crash site, where they worked until lunchtime. After a quick lunch they resumed work until 4:30 P.M., then were transported back to the suiting-up facility, where they were decontaminated. They returned to their hotel to shower, then usually back to the command center by 5:30 in order to report on their day's activity to the other members of the group. Later in the evening they ate, sometimes at a local restaurant, but until bedtime they were in and out of the command center for progress meetings, group planning meetings, and talk and more talk with their colleagues. Sometimes this meant everyone talking to everyone—whatever it took to sort through and organize the information.

Cox found that he got to bed about 1:00 A.M., if he was lucky. After a few days, it was hardly surprising that emotions rose to the surface. "The tears are going to come, almost nobody gets through it without that and it's in no way a sign of weakness, it's a sign of being human," Cox explained. "You cannot work in that level of carnage and not be affected by it."

The tail was not just the only immediately recognizable part of the wreckage—it also yielded the least-damaged hydraulic components, the main rudder power control unit (main PCU) and the standby rudder power control unit (standby PCU). They were carefully removed under the supervision of personnel from Parker Hannifin, the company that manufactured the main PCU and an official party to the investigation. The objective was to try to take them out of the wreckage without altering their settings. Rudder movement was not recorded on the flight data recorder, so it was vital to preserve any indication of which way the rudder might have been turned at the time of the crash, if that were possible. A similar procedure was followed with the other actuators, the hydraulic levers and rods that extend flaps and slats or move ailerons and elevators up and down. First, they were carefully measured and photographed. Then they were secured with wire so that their settings would not be altered as they were removed from the wreckage.

The care and perseverance paid off. By measuring the extension of various actuators, the Systems Group could reconstruct their positions on the wings or tail. The extension of the flap actuators on the left wing, for example, corresponded exactly to where they would have been if they were extended to the Flaps 1 position, appropriate for the speed of the aircraft just prior to the loss of control. The flap actuators on the right wing told a similar story, as did other wing extensions at the leading edge, devices known as Kreuger flaps, which help accelerate the airflow over a slow-moving wing. Examination ruled out the possibility that the flaps and other controls had been asymmetric, a situation that might have caused the aircraft to lose lift on one wing and lead to a stall. The elevators were found set to a position that normally would have trimmed the aircraft to a 14° nose-up position.

The main rudder PCU had 2.38 inches of actuator rod protruding from it, but the rod was bent from impact damage. After noting its position, Phillips carefully moved the input arm of the standby rudder PCU to see if its bearing rotated. If it failed to move, the investigators would have suspected that it might have seized due to galling, a situation that had been discovered on the Flight 585 standby rudder PCU at Colorado Springs. But the bearing rotated. It had not seized.

Having recovered as many flight controls and their power units as he could from the wreckage, Phillips declared Phase One of his investigation at an end. Phase Two, the minute, microscopic examination of what had been found, and, where possible, its reconstruction and testing, was about to begin. Phillips and other members of the Systems Group left Pittsburgh for Seattle on September 19 and went to Boeing's equipment quality assurance facility at Renton, Washington. Awaiting them were components from the crashed plane that had been sealed and dispatched ahead of them under tight custody procedures, not dissimilar to the way police at a crime scene handle evidence being gathered for forensic examination. There had been acute embarrassment earlier when several small parts of the main rudder PCU from Flight 585 had gone missing after being sent from United's San Francisco facility to Parker Hannifin for examination. A full investigation into the affair had not solved the mystery. This time, Phillips hand-carried some parts with him to Renton.

There, guided by Boeing engineers with intimate knowledge of the parts, they spent a day examining the main rudder PCU, the rudder trim actuator, and the standby rudder PCU. The rudder trim actuator is an electrically-controlled device that puts a slight offset into the rudder whenever, for example, one engine is developing less power than another, and a greater offset if one engine fails. Although the examination was cursory compared to some of the tests they would perform later, nothing appeared to be amiss. Everything was carefully measured, described, and even x-rayed. Because it had a bent actuator rod, there was a limit to the tests that could be performed on the main rudder PCU. The standby rudder PCU, designed as a backup to the main rudder operating mechanism, was rarely used in practice, and this one showed no abnormal symptoms. When a quantity of hydraulic fluid was discovered in the unit, Phillips ordered a sample dispatched to Monsanto, the fluid's manufacturer. Hydraulic fluid can become contaminated with microscopic metal particles, with water, even with air bubbles, all of which can affect its performance. In the Colorado Springs crash, all the fluid had been boiled off by the fire, but in the wreck of 427, the tail where the unit was installed had remained relatively cool.

The following day, the Systems Group was on the move again—to Parker Hannifin's large plant in Irvine, California, south of Los Angeles. The company specialized in finely turned, precisely measured engineering components for applications as diverse as amusement park rides and snowmobiles, wheel and brake systems on single-engine Cessnas, and sophisticated flight systems aboard airliners, executive jets, and military jet fighters and helicopters. But airplane hydraulic systems were its forte.

There are three ways of making a flight control surface move, be it an aileron, an elevator, a flap, or the rudder. One is by a system of cables and pulleys directly connected to the part, as on small single-engine Cessnas and Pipers. Such a system was a backup for the Boeing 737 ailerons. Another is by signaling an electric motor directly connected to the part it moves. And the third is hydraulics.

Such systems can be found on the automobile. The parking brake is operated by a cable. The wipers, the sunroof, the windows, and the adjustable driver's seat are all powered by small electric motors. They are cheap and reasonably powerful, but they have some drawbacks. Inexpensive motors lack the ability to make precise, accurate movements, and more expensive motors that can be finely controlled are usually heavier. And electric motors require servicing or replacement: carbon "brushes," which convey electric current to the spinning commutator, wear down and need replacing; bearings need lubrication and eventually fail. Automobiles use hydraulic systems in power steering, automatic transmissions, and brakes. Because their moving parts are immersed in oil, hydraulic systems suffer less wear. They are capable of transmitting powerful movement in a very controlled fashion. The backhoe digging a trench at a construction site employs hydraulics both to plow through hard earth and to precisely place the contents of its scoop into the bed of a waiting truck.

In an automobile, pressing the brake pedal moves a piston in a servo unit—a device that determines the force and direction of a gas or fluid in a pneumatic or hydraulic system. In the hydraulic system used in most cars and small trucks, fluid is forced down four narrow-diameter tubes, one leading to each wheel. At the wheel, the fluid enters another servo unit and pushes a piston that forces the brake pads against a brake drum or disc attached to the wheel, which slows down the car. Think of an auto moving at 60 mph that is brought to an emergency stop in seconds. It requires very little effort by the driver to overcome a large amount of kinetic energy, thanks to the fact that the fluid in the braking system is already under pressure. The energy to maintain the pressure on the fluid in a car's braking system comes from the engine. That same energy, however, could equally be provided by a pump, like the one that powers the hydraulics in the steering system.

Aboard the Boeing 737 are five separate pumps, each capable of powering a separate hydraulic system. Two are mechanical pumps driven by the engines and backed up by two electrical pumps. The fifth is an electrical pump driven by the aircraft's main electricity supply. Each pump delivers about 3,500 pounds of pressure. This may seem a like lot until you realize that the control surfaces on a large airplane must act against the rapid flow of air rushing at hundreds of miles per hour over the wings or past the tailfin. Each control has what is known as its blowdown limit, the extent to which it can be deployed before the hydraulics are defeated by the airflow. Although the rudder can move 26° in each direction when stationary on the ground, speeding through the air it can move much less, and 15° would be a major deflection during flight. In any event, the greater the airspeed, the more effect a given rudder deflection has on the plane, and the greater the hydraulic power needed to make that deflection. Moreover, planes generally have tremendous power in their hydraulic systems in case extra pressure is needed to produce what is called a breakout force—a force great enough to overcome any obstruction that might have jammed one of the moving parts in the system.

Parker Hannifin is a leader in US hydraulics, but when the NTSB Systems Group trooped into one of its labs two days after leaving Seattle, the people working there seemed on the defensive. The NTSB team's mission was to fully test the main rudder power control unit. Compared to the last Boeing 737 PCU Greg Phillips had examined, from Flight 585, this one was in reasonably good condition and might be capable of being pressurized to test its operation. The NTSB had made no official announcements, but word was already filtering out through the various groups that analysis of the flight data recorder suggested that the rudder might have played a role in pushing the aircraft into its fatal roll.

Since the PCU was the main steering force in the back of the plane, it was subjected to the most detailed scrutiny the team could give it. First they examined its external physical appearance. Although some linkages on the outside were damaged, the device was basically in good condition. The hydraulic fluid that remained in the component was, along with the fluid filters, extracted and placed in sealed containers for testing by Monsanto, the fluid's manufacturer. The main actuator piston, the unit that actually moves the rudder assembly, was removed, and a borescope was inserted into the PCU to enable investigators to peer inside. The scope had a sort of low-powered microscope with a prism at one end that could be swiveled to examine the interior of the PCU. The investigators discovered that the chamber was clean and free of any marks that indicated jamming. The rod leading from the piston was bent in the crash and was replaced with a new one. The team then performed a range of tests to gauge the freedom of movement of the PCU's parts. These moved effortlessly, as they were designed to, belying the fact that they had come to earth aboard an airplane traveling nose first at more than 300 mph.

The rudder PCU is an extraordinarily complicated piece of machinery. There is no easy way to describe it in detail. Put simply, it takes commands through cables from the cockpit and through separate electrical signals from the yaw damper, and triggers the hydraulics to push the rudder left or right, quickly or slowly. The way in which these commands are executed is far from straightforward. Americans would call it a kind of Rube Goldberg mechanism, and the English might describe it as a Heath Robinson sort of affair, after the famous cartoonists whose drawings of complex arrangements of ropes and pulleys and gears and levers were ingeniously arranged to perform what were, in the end, relatively simple tasks. Rudder cables travel from the pedals at each pilot's feet through a series of guide pulleys beneath the floor of the cabin to the tail, where they are connected to a series of bars, rods, and so-called quadrants and torque tubes. To operate the rudder, a pilot presses a left or a right pedal, and the system transmits the pilot's command, through a series of connecting rods and linkages, to the hydraulic system in the tail. The final agent of the command, the device that actually moves the rudder, will be either the main rudder PCU or the standby system.

The main rudder PCU is powered by two main hydraulic systems, A and B, with A being the default, "always on" system. The standby rudder PCU, a separate hydraulic system, is switched on when both A and B have failed. It has its own power unit and actuator and, apart from the rudder pedals, cables and torque tube conveying commands, has nothing in common with the main power control unit. The standby unit is larger and is manufactured by a different company, Hydraulic Units, Inc. When it is not in use, which is most of the time, the standby unit is unpressurized. Every time the rudder is operated, the standby input rod moves back and forth uselessly. During the Flight 585 investigation, it was thought that galling, which occurred on the unit's bearing because of forceful rubbing by the output arm, might have caused the rudder to jam.

The designers of the Boeing 737 rudder were playing catch-up with McDonnell Douglas. The 737 was to compete with the DC-9, and the DC-9 was already flying. Most of the rudder components were taken from the Boeing 727. As the designers worked, they had to bear in mind the redundancy requirements of the FAA, which had to certify the design. The airplane had to be capable of being operated safely and completing its journey when almost anything essential for flight jammed or failed (assuming that the wings and the tail would always remain attached). It had to be capable of flying on a single engine, for example. The aileron system of hinged surfaces on the trailing edge of the wing had backup in the form of the inboard spoilers, four on each wing, which could play a similar role should the ailerons fail. (A spoiler is a hinged surface on the upper part of the wing that increases drag and reduces, or "spoils," lift by disrupting the airflow.) Deploying spoilers on one wing in place of ailerons causes that wing to drop, and the aircraft will turn in that direction. On the tail, the stabilizer system, which angles the rear "wings" up or down to change their angle to the airflow, could fulfill the same role as the elevators, which are hinged surfaces on the trailing edge that push the nose up or down for climbing or descending.

There was only one rudder on the 737, an uncommonly large creation because, unlike the 727 that inspired it and that had all of its engines mounted on the rear fuselage, the 737's engines were wing-mounted. If one engine failed, especially during takeoff, the leverage produced by the remaining engine on the other wing would push the aircraft's nose around in a severe yaw and lead to a loss of control. Conventional rudder systems could not correct this. A larger rudder was necessary to counteract the yawing during an engine outage. But there were problems with this approach, and they had to be solved if the FAA was to be kept happy. Other aircraft fulfilled the FAA's redundancy requirements by having a split-panel rudder operated by separate PCUs, in effect one small rudder atop another. If one failed, the other could fill the breach; if one jammed, the other could cancel it out. Boeing responded with a single rudder, but with split controls. If the main PCU failed, the standby PCU could be used. If the main PCU jammed, there was sufficient power in the standby PCU to perform a "breakout" from the jam. Boeing engineers took the design a stage further. The valves in the main PCU that directed the flow of hydraulic force were duplicated, one inside the other. If one jammed for whatever reason, the other could carry on.

Boeing's designers faced two choices as they were selecting a design for the unit to move the rudder. One option was two power units, one for each direction of travel, left and right. Another was to combine both directions of travel in one power unit, a device with an actuator that could both push and pull the rudder. The latter was the more elegant solution, but it required extensive designing to keep it from becoming an unwieldy object, not much smaller or lighter than two units. Boeing's solution was the dual concentric servo valve, a soda can-sized component connected to the pilot's pedals via the input lever, the torque tube, the aft quadrant and, finally, the rudder cables. The servo valve was essentially a hydraulic switch and formed the heart of the PCU. This particular servo valve was especially sophisticated in the sense that it could trigger a powerful flow of hydraulic fluid in *two* directions. One direction, as commanded by the right rudder pedal, directed the flow to turn the rudder right, and the other direction turned the rudder left.

If that were not complicated enough, more was to follow. Two hydraulic systems, A and B, had to be accommodated and be capable of powering the unit together or separately. Mechanisms had to be put in place to prevent the unit from jamming: as hydraulic fluid is injected into the actuator piston chamber to force the piston in one direction, fluid has to be rapidly evacuated from the other end of the chamber to allow for its movement.

And what if the servo valve jammed, causing hydraulic fluid to inadvertently either push or pull the rudder actuator? That could produce what pilots call a hardover, a situation where the rudder is permanently pushed over to the left or the right, making the aircraft difficult to control. Boeing's solution was to design the servo valve with two valves (called slides) working together, one within the other. One slide would take over if the first one jammed, and vice versa. The slides are two narrow, tightly fitting cylinders that can slide back and forth in the servo valve.The outer slide is hollow and the inner one slides to and fro within it. The outer one has holes bored through it that can match up with circumferential grooves in the inner one. The chamber in which they both move also has holes in its wall. Hydraulic fluid under pressure can pass in and out of these holes. The holes are present on opposite sides of the chamber and are critical to the mechanism of the servo valve, and ultimately the entire PCU.

The servo valve allows hydraulic fluid to pass from one side of the chamber to the other, or not, as the case might be. Think of a light switch that allows power to pass from one terminal to another and then light a bulb. When the rudder is in neutral, not being operated, the slides block up the holes in the walls of the chamber and fluid cannot pass from one side to the other. The switch is off. Power cannot pass through. Move the slides slightly, and holes and grooves in them line up with the holes in the chamber wall. Hydraulic fluid under pressure gushes through and out the other side. The switch is on. Power is moving through. The fluid passes on, enters the actuator chamber, and moves the piston that is connected to the rudder. The rudder moves.

Now think of a more complicated switch like that found on a power drill. Turn it one way and the drill turns clockwise. Turn it the other way and the drill turns counterclockwise. When the slides in the servo valve line up with one set of holes, fluid can rush to the aft chamber and the rudder piston is pulled. When they line up with a different set of holes, fluid is diverted to the forward chamber and the rudder piston is pushed. Surprisingly, the total movement of each of the slides within the servo valve is just 40 thousandths of an inch—the thickness of a dime.

But this is only half the story. When fluid under pressure is allowed to push the rudder piston, an equal amount of fluid must be allowed to make the return journey—otherwise the piston will jam. So, two sets of holes are exposed for each left or right command. One set moves the rudder and the other set allows displaced fluid to return to the main hydraulic system.

Greg Phillips understood the workings of the servo valve as well as anybody at Boeing or Parker Hannifin, which manufactured it to Boeing's patented design. Following the so-called Mack Moore incident in Chicago more than two years earlier (in July 1992), Phillips had a good idea of one way the PCU could malfunction. Some of the units were microscopically defective, allowing the slides to travel just a little further than normal, which in turn allowed fluid to travel through the wrong holes. The jam reported by Captain Moore could have been caused by this defect, which would block the return flow. The reversal reported by the United mechanics when they tested the unit from Moore's aircraft was something else and required additional explaining.

As he continued his examination of the rudder equipment from Flight 427, Phillips found himself with feelings of *déjà vu*. He'd been here once before, in this very building, chasing a hydraulic will-o'-the-wisp. Something had been unearthed that, if it could occur in real life in the back of an airplane and not just artificially induced on a lab bench, might explain some of the strange things that had been happening to 737 aircraft, and might even explain the crash of Flight 427.

It had been the subject of a safety alert, and Parker Hannifin was at that very moment working on a backlog of orders to produce redesigned replacement PCUs manufactured to prevent the overtravel that caused the Mack Moore incident. Phillips's colleague, Hector Casanova, who had been plowing through the maintenance work records from the USAir shop, had confirmed that Flight 427 had a new PCU installed because the old one was leaking. But that was in February 1993, two months before Boeing issued a service bulletin on the PCU, and a full year before the FAA published its Airworthiness Directive mandating replacement of the PCU within five years. None of the new, redesigned units were yet available. In fact, they hadn't even been designed. Flight 427's PCU had been replaced with a new unit just like the one pulled off its tail.

The Systems Group pressurized the PCU with the aid of Parker Hannifin's equipment and ran some more tests. Vigorous strokes of the input levers were made to see if it would jam or reverse. It functioned perfectly and smoothly. Electrical connections were made to simulate the effect of yaw damper commands. It passed all of the yaw damper tests, except one, a simulation designed to measure the rate of travel of the servo-valve slides. The slides were removed from the servo valve and examined beneath a microscope.

Nothing except normal wear was found. The servo valve was reassembled and tested separately; it passed all of its important tests. It was time to go, and Phillips stripped any new parts from the PCU to restore it to its post-crash condition. It was then packaged and handed over to an NTSB staffer, continuing the chain-of-custody procedure that had brought it there in the first place.

Two weeks later, on October 3, 1994, Phillips turned up at Boeing's Renton plant again at the head of his group. Irrespective of his growing suspicion that the rudder was implicated in the crash, he was determined not to be the "guy with a theory." He was going to lead a full review of all flight control systems. For the next five days, they sorted through a small pile of actuators and attachments to various control surfaces, all showing some signs of being burnt and battered from the crash. Although some were badly damaged on impact and incapable of being tested, nothing unusual was found. The cockpit dials that displayed hydraulic pressure for the A and B systems were also found. The needles were missing but there were tiny traces of white paint on the face of the dials. They were witness marks from the painted needles—paint had ghost-marked the dial face as the needles flicked forward on impact as the plane crashed. They were positioned at about the 3,000-pound pressure mark for each system, a normal operational level. Unlike the DC-10 at Sioux City, this plane appeared to have had working hydraulics.

There was a further breakthrough when they sorted through the pieces of the pilot's and co-pilot's control columns. The control column, or "stick," is used to operate the elevators and the ailerons, causing the aircraft to climb, descend, or to turn left or right in a gentle bank. It is hinged at the base, and by pushing or pulling it, the nose is put up or down and the aircraft climbs or descends. There is a control called a wheel atop each column that looks like a small cutaway steering wheel. It turns a small drum on which cables that extend to the aileron actuators are wound. As one wheel is turned by the pilot flying the aircraft, the other pilot's wheel also turns. When Phillips and his team painstakingly reassembled all the battered pieces of the two columns, they realized that some moving fittings, like the flattened hub of Captain Germano's wheel and the cable remaining in the badly damaged cable drum of the co-pilot's wheel, suggested that the wheels had been turned to the right at a 40° angle. This was a major breakthrough.

The Systems Group had earlier determined the elevators were in a 14° nose-up position; now came confirmation of the position of the ailerons. Another piece of the jigsaw had been put in place. Greg Phillips now knew for sure two things the pilots were doing in their final moments. They were wrestling with the wheel and trying to get the nose up. Those two pieces of information would help him move closer to deciding if any other control might be somehow responsible for the crash. But only a bit closer.

Chapter 10

SECRETS OF THE FLIGHT DATA RECORDER

Dave King, the veteran British air crash investigator, made one criticism in his report to Cynthia Keegan on Flight 427. He called the state of the regulations that allowed US airlines to continue flying with a minimum standard for flight data recorders "a handicap to the investigation." He added, "The addition of flight control positions"—routine parameters for the flight data recorders used by airlines outside the US—"might have enabled the investigators to eliminate certain possible causes, if not identify the cause of the flight path departure."

Because the flight data recorder aboard Flight 427 recorded just a handful of parameters, Tom Jacky's unenviable job was even less enviable. He had led the Flight Data Recorder Group investigating the crash of United Flight 585 at Colorado Springs and now led the Aircraft Performance Group for the Flight 427 probe. First, he had to reconstruct the final flight path of USAir 427, then decide what combination of rudder, aileron, or elevator—deliberately or accidentally applied—could have caused a 60-ton aircraft to suddenly spin out of the sky and crash before its pilots had the chance to snap off an intelligible Mayday.

Jacky, who had joined the NTSB as an engineering programmer shortly after graduating as an aeronautical engineer, was steeped in air accident investigation—he'd been with the NTSB for nine years. But he soon realized he had been handed a near-impossible task by Tom Hauteur. A similar exercise in "reconstruction," one of the first ever attempted, had been tried in the wake of the crash of Flight 585 but had proved inconclusive. Admittedly, at Pittsburgh he did have a little more to work with. Radar coverage of the doomed Flight 427 was better than for Flight 585, and a few more flight parameters had been wired into the flight data recorder. There was, for example, more data on what the engines were doing. The 427 FDR had also recorded control column positions, if only in the fore and aft directions, which meant he could determine whether the pilots were attempting to push the aircraft's nose up or down. Then again, there were critical parameters that were not available. The motion of the control wheel, which operates the ailerons on the wings to induce roll, was not recorded. This device works like an automobile's steering wheel:

swing the wheel to the left and the aircraft's left wing dips slightly and the plane turns to the left. Swing it to the right and the opposite happens.

Jacky would have liked more wind measurements as well, but he could assume that the weather had been relatively settled. At Pittsburgh, there were none of the unpredictable Rocky Mountain updrafts, downdrafts, and rotors that so complicated the Flight 585 investigation. Flight 427 was flying in a stable high-pressure zone across a relatively low-lying landscape that offered few obstacles to the light breeze.

Jeremy Akel, one of the NTSB's flight data recorder specialists, had already done the grunt work on the flight data recorder, similar to what Jacky had done for Flight 585. The recorder—a Loral-Fairchild Data Systems model F100—had been retrieved from the wreckage the evening of the crash and sent to Washington for analysis. The exterior casing was badly damaged, as was the interior, where dirt and aircraft insulation had gotten in. The solid-state memory unit was torn from its mounting and lodged at the rear of the case. The crash had destroyed many of the recorder's circuit boards, but the memory module was in surprisingly good condition. The recorder was attached by cable to another recorder in the NTSB lab, and its last 10.5 hours of flight data were downloaded for analysis.

Because of certain peculiarities in this model of flight data recorder, data could only be accurately recovered in 4-second segments. This meant that the final 1.9 seconds of the recording was corrupted. Akel ordered a copy of the entire recording to be sent to Loral-Fairchild for reconstruction of the missing segment, and within a few days, he had enough information to reconstruct the plane's final 30 seconds.

Of course this meant only that he could say roughly what the airplane did; he could not deduce why. In addition to the heading, altitude, airspeed, and vertical acceleration data (which was all that had been recorded for Flight 585), the recorder on Flight 427 had the parameters of engine output, engine exhaust temperature, fuel flow, pitch and roll attitude, and fore and aft control column position. These added little to the investigators' knowledge of the crash, but they were better than nothing.

All told, the data showed an aircraft that was completing a turn to the right, its wings rolling back to level flight. Then it started to roll violently to the left, and its nose also started to turn left. Soon, the aircraft had rolled almost onto its side. And as the rolling motion increased, its nose started to point downward. Some 20 seconds after the incident began, Flight 427 was upside down and twisting toward the ground. In the next 10 seconds, it completed a full 360° roll, still pointing downward. After it rolled 90° more, the rolling momentarily halted, the nose came back up slightly, and the wings jerked in the opposite direction. But then the relentless left roll and corresponding dive resumed, until impact a second and a half later.

Apart from the control column position, nothing in the data extracted by Akel told him anything about the pilots' actions or what other controls they were using. The aircraft's motion had all the hallmarks of a rudder-induced roll and dive—no other control could have had such dramatic effect. But Akel had no data for what the rudder pedals, or indeed the rudder itself, were doing. Neither did he know what the ailerons, the control surfaces on the wings, were doing. The data for the control column position provided a little bit of suggestive information—the numbers suggested that the doomed co-pilot was hauling back on the stick in a vain attempt to bring the aircraft's nose up. But other than that, the roll and dive were a mystery.

At Hauteur's urging, Akel had embarked on a study of Boeing 737 incidents, and it was at his suggestion that Systems Group head Greg Phillips had begun compiling a list of mysterious rolling or yawing incidents involving the 737. Since the crash of United 585 at Colorado Springs, in March 1991, until June 1994—3 months before the crash of Flight 427—Phillips had tallied 13 incidents, or an average of one incident approximately every three months. The first item described unexplained rudder movements two days apart in February 1991 aboard a United Airlines 737. A week afterward, the very same aircraft crashed as Flight 585. The second item on Phillips's list was the so-called Mack Moore incident, when a United Airlines captain felt his rudder pedals jam as he was testing his aircraft prior to takeoff in July 1992. That event led to the discovery of a fault in the design of the rudder power control unit on the 737 that could cause the rudder to flip to the wrong side.

Other incidents followed, including two other rudder jams encountered on the ground. In April 1993, the rudder on an Air New Zealand 737 started moving left and right every 2 or 3 minutes in flight at an altitude of 33,000 feet. The rudder pedals also moved. The autopilot was able to counteract the effect on the aircraft by using the ailerons. No cause was ever found for these anomalies. In July and August 1993, Air France reported two "violent" yaws. Also in August 1993, a Southwest Airlines 737 experienced a midair rudder hardover that injured a flight attendant. Ironically, it was the aircraft's first flight after mechanics replaced a rudder power control unit that malfunctioned during a ground test. A Philippine Airlines rudder malfunction in November 1993 was potentially more serious because it occurred during takeoff, which was aborted. The captain reported the aircraft suddenly swerved to the left when traveling down the runway at 80 knots. The uncommanded left rudder application aboard an America West 737 descending to Las Vegas in late March 1994 was blamed on a faulty electrical part, a solenoid. A similar event in May was the second such incident to hit America West that year and the seventh overall, resulting in the removal of yaw damper components from its fleet.

Akel also ran down five serious departures from controlled flight that initially appeared to bear some similarity to the last seconds of Flight 427, not all of which had made it onto Phillips's list. Three had resulted in fatal crashes; in the other two, the pilots regained control and landed safely.

The nonfatal list included a United Airlines 737-500, a newer design incorporating a shorter fuselage than the 200 or 300 series but using the exact same wings and control systems. That aircraft had misbehaved at Denver, Colorado, in April 1993 and afterward landed safely. The other nonfatal incident that was on Phillips's list was a Continental Airlines 737-300 on a flight from Tegucigalpa, Honduras, on April 11, 1994. Its pilots landed in San Pedro Sula, also in Honduras, after reporting that they almost lost control cruising at 37,000 feet at an airspeed of 242 knots. The airplane had suddenly rolled 9° to the left and yawed nose left at the same time. Boeing afterward declared the cause had been a faulty yaw damper that pushed the rudder to 2.5° left and held it there. It was, Boeing said, the maximum deflection permitted for the yaw damper. The crew responded by controlling the aircraft with the opposite aileron and disengaging the autopilot, although they left the yaw damper switched on.

The movements of the plane in the Continental incident, rated as serious by the airline, read as an almost straight line compared to the jagged, disturbed graph corresponding to the movements of Flight 427. The Continental plane lost just a couple of hundred feet in altitude, compared to almost 5,000 feet by Flight 427. Maximum acceleration was less than 0.5 G, compared to almost 4 G for USAir 427. Left and right heading changes were 2° or 3° of the compass per second for the Continental flight compared to much more rapid heading changes for USAir 427. (At one stage, USAir 427's nose shot through 30° in just 1 second.) And while the Continental incident produced some very uncomfortable pitch oscillations (nose up and down), they never exceeded 2° or 3° per second. The graph for the roll rate told a similar story: while the Continental aircraft's wings rocked alarmingly from one side to the other, at worst these movements never exceeded about 5° per second. USAir 427's wings, on the other hand, at one stage slammed from 50° down on one side to 50° down on the other side in less than 3 seconds, a rate exceeding 30° a second.

The fatal incidents on the list included the June 6, 1992, crash of the Copa Airlines 737 in Panama, on which Hauteur had worked more than two years earlier. There was also the March 1994 crash of a Sahara Airlines 737, a training flight at New Delhi airport that crashed shortly after taking off with a simulated engine failure. Finally, there was the crash of United 585 in Colorado Springs, one of the very few accidents that the NTSB had failed to explain.

The Sahara aircraft went through some extremes of flight comparable to Flight 427, but they exhibited a different pattern. The last moments of the Copa crash, before the aircraft broke up in mid-air, bore a closer similarity to Flight 427 in many respects. Its rolls, accelerations, and heading changes were of a similar scale, although the Copa aircraft's pitch oscillations were nowhere near as dramatic as those of Flight 427.

But it was Flight 585 that bore the greatest similarity to Flight 427 in numerous respects, despite the fact that its uncontrolled flight was briefer because the aircraft was closer to the ground, and the pitch and roll data for 585 was not recorded and therefore had to be derived by other means. The similarities in the two crashes could have marked an important breakthrough in the Flight 427 investigation—if only the NTSB had a definitive cause of the crash of Flight 585. Now it looked as though the NTSB had two similar, and similarly unsolved, accidents on its books.

Jeremy Akel signed off his ancillary flight data recorder report comparing the various incidents and accidents on December 19, 1994. The following day, near Raleigh, North Carolina, there was another incident aboard a USAir 737. After experiencing a sharp uncommanded rudder movement, the aircraft yawed to the left. The pilots switched off the autopilot and autothrottle, and the event, which lasted about 2 seconds, ceased. They reported that at no time did the rudder pedals at their feet seem to move. Afterward, mechanics found an electrical fault in part of the autopilot system and fixed it. However, it emerged that a month earlier the same aircraft had a sudden uncommanded rudder movement while at 5,000 feet and coming in for a landing at Philadelphia. This time, the pilots did feel the pedals move. When they switched off the yaw damper, the flight returned to normal. There were two mysterious aspects to the incidents. First, the autopilot on a Boeing 737 only has authority over the ailerons, the stabilizer, and the elevators. It never commands the rudder. And second, when the yaw damper operates the rudder, it does so completely independently of the pilots' controls. They should not feel the rudder pedals at their feet move.

Tom Jacky fleshed out Akel's carefully compiled data and tried to fit it into the context of the air traffic above Aliquippa and on the approaches to Pittsburgh International Airport. He obtained the radar data that NTSB air traffic control specialist Alan Lebo had used for his investigation and plotted them on a US Geological Survey map of the accident area. Then he began looking for further clues in the neighborhood of the crash.

Lebo had looked for heavy aircraft whose wake may have affected Flight 427, and found none. Jacky decided to look at any aircraft, heavy or light, that had been in the vicinity of Flight 427. There was the possibility that Flight 427 might have swerved precipitously to avoid a midair collision with an another aircraft that had gone unnoticed and unreported by the controllers. Another possibility was a loss of control caused by wake turbulence. Flight 427 was approaching Pittsburgh at the tail end of a busy period, and there were two aircraft in its neighborhood. One, a commuter flight, had crossed Flight 427's intended track at right angles about 3 miles ahead and 1,200 feet below, and was unlikely to have been a factor. But there had been a second aircraft, a Delta Airlines Boeing 727 also en route to Pittsburgh. It was some 4 miles ahead and had been routed along the same track that Flight 427 was maneuvering to join when it went out of control. Jacky asked for the radar returns of the Delta plane, Flight 1083, and laid them over Flight 427's returns on the Geological Survey map. As Lebo had predicted, they crossed at almost the exact point in the sky where Flight 427 had started to go out of control.

The Boeing 727 was the aircraft from which the 737 had been derived in Boeing's race to produce a competitor to McDonnell Douglas's DC-9, and the two Boeing planes had much in common in the fuselage and wing design. However, the 727 had three engines, all grouped together at the tail, while the 737 had two engines slung beneath the wings. The 727 was not considered an aerodynamically "dirty" aircraft, at least as far as wake turbulence was concerned. Further inquiries revealed that the aircraft was being flown in a "clean" condition, with flaps extended to the Flaps 1 position about the time it would have passed through the area of sky later occupied by Flight 427. Still, it was too much of a coincidence to be ignored.

Very little was known about wake turbulence until the FAA and NASA devised a series of tests over 30 years ago. In order to demonstrate the phenomenon, smoke generators were mounted on tall pylons, and various aircraft were flown through the plumes of smoke. The aircraft whipped the smoke into characteristic spirals. Researchers found that the worst turbulence came not from the airplane's engines, as was once suspected, but from its wingtips. Some of the air that flows over the wings moves slantwise toward the wingtips, then spins off in a vortex, or miniature horizontal whirlwind. Each of the two vortices, one for each wingtip, rotates in opposite directions.

Researchers found that while the vortices sank vertically through the air, in calm air they could persist at or near the altitude at which they were generated for 2 or 3 minutes. They could also be propelled to one side by a breeze.

Following the May 1972 crash of a DC-9 at Fort Worth, which was blamed on wake turbulence from a DC-10, the FAA performed a series of tests. These showed that an aircraft descending from the right side into the wake vortex from the left wing of a preceding heavy aircraft first experiences a strong sideways blast of air from the upper side of the vortex. Pilots were found to have a tendency to overcompensate with the ailerons to bring the right wing back to the level position. Then, as the aircraft passed through the core of the vortex, the right wing entered the opposite side of the rotating air, which now forced that wing down. If the right aileron was still applied, this could cause the right wing to be forced down even more suddenly, turning the aircraft onto its side. Furthermore, research at McDonnell Douglas, manufacturer of the DC-9 and DC-10 aircraft, suggested the rolling power of a DC-10 vortex could exceed the ability of a DC-9's aileron to respond.

After discussing his findings with the other experts in his group, Jacky secured Boeing's permission to use its simulator in Seattle for a series of tests. Boeing's delegate on the Performance Group was James Kerrigan, the lead engineer on the aerodynamic stability and control group for the 707, 727, and 737 aircraft. After almost 30 years of work on the 737, he was one of the company's most knowledgeable engineers, and there was little he did not know about the 737's flight control surfaces, the movable panels on the wings and tail that controlled its movement through the air. He arranged for other Boeing experts, including senior test pilot Mike Carriker, to join him and Jacky. Also present were the NTSB's John Clark, who had attempted a similar exercise in the Flight 585 investigation, and two experts from NASA, including wake turbulence specialist George Green.

Just eighteen days after the accident, on September 22, 1994, in Boeing's M-Cab simulator in Seattle, Jacky and his colleagues attempted to replicate Flight 427's final passage through the air by using different control commands. First, they ran a series of data collecting, or "what if" tests, manipulating engines, rudders, flaps, and so on. The remainder of the day was spent putting the simulator through 45 different scenarios to see if any of them resulted in a "flight" that matched any of the final seconds of Flight 427. Nothing conclusive emerged from the tests.

Three weeks later, Jacky and his colleagues returned to Seattle. This time, the simulator was programmed to replicate the effects of Flight 427 entering the wake vortices of an airplane on the same course some 4 miles ahead. Only one computer program was available to simulate wake vortices, and that was the program developed three years earlier to simulate the impact of a rotor, a much larger vortex, on Flight 585 at Colorado Springs. It was a useful program, however, and had the capability to vary the strength and diameter of the rotor's core, as well as changing its rotation. Two counter-rotating rotors were inserted into the program to represent wake vortices from the Delta 727. Core sizes were varied between 4 feet and 17 feet, and separated in the simulator by 85 feet to represent the wingspan of a 727. Vortices are invisible in the real world, but pilots flying the simulator could see the vortices represented by two cylinders joined by a red line on the cockpit window display. Boeing 737 pilots from the various parties flew the simulator and agreed the new package gave a realistic response. They flew the simulator into the wake vortices at varying angles. Invariably, the pilots were easily able to recover control after an initial roll. Either the wake vortices from a 727 had little effect on a 737, or else there were serious inaccuracies in the model programmed into the simulator.

When Boeing is flight-testing a new aircraft, it repeatedly measures at a very rapid rate—in increments as small as 20 times per second—how the components of the plane are working together to produce controlled and predictable flight. How far the nose moves in a turn, or a wing drops in a bank, or the rate at which the aircraft climbs or descends, or its acceleration, or any one of dozens of parameters are measured almost constantly by sensors placed all over the aircraft. After the flight test, Boeing engineers can download the data from a sophisticated quick-access flight data recorder and discover what the aircraft was doing at any particular moment. They can see which flight control surface, like an aileron or rudder, was deployed and how much force the pilot needed to apply to move it. Boeing engineers can even measure the slippage of an aircraft skidding sideways through the air (sideslip) as it makes a tight turn, just as a race car skids from the center to the side of the track at a curve. On a test aircraft set up to record 60 parameters, the minimum number Boeing engineers might consider worthwhile, they would have sensors providing 72,000 data points per minute. And if the number of parameters is raised to 100, which would not be unusual, they would have 120,000 data points per minute to work with.

But aboard a standard production Boeing 737-300 aircraft of Flight 427's vintage, only six parameters that describe the airplane's motion are sampled on the flight data recorder, and each only once a second. This meant that Tom Jacky knew just six things about the aircraft's passage through the sky for each of the roughly 30 seconds during which it spun out of control and crashed. That came to some 180 data points. It was more than Jacky had seen for flight 585, but the problem was, did this amount to any meaningful kind of data, and could he rely on it?

In Seattle, Harry Dellicker was facing a similar conundrum. An engineer in the company's aerodynamics group, he had been with Boeing for 17 years, 12 of them spent coming up with better ways of analyzing the reams of data from the test flights of new airplane designs. One of his specialties was kinematics, the study of the mechanics of motion. Dellicker plotted velocity and acceleration and fit curves to the data points; since he was dealing with highly instrumented test aircraft, he could cross-check one parameter against another in order to test the accuracy of the data recorded in flight. For example, if a printout after a test flight showed that the rate of acceleration increased over a time interval in which the airspeed decreased, you could assume that one measurement or the other was inaccurate. Or if the airplane was in a tight turn and yet there was zero sideslip, something else would need fixing.

As every high school math student knows, Dellicker liked to point out, if you know the speed of an auto on a straight, level highway at various points in time, you can calculate the distance it has traveled, plus its acceleration or deceleration. If you know the weight of the vehicle, you can use the equation *Force = Mass × Acceleration* to work out the engine force needed to produce its acceleration, or the braking force required to slow or stop it.

Things are more complicated with an airplane. Whereas a car restricted to a long, straight highway can move in two dimensions only, an airplane has six three-dimensional options. Three of these are the getting-from-A-to-B, or translational, modes: vertical (climbing and descending), lateral (going left and right), and longitudinal (flying straight and level). The three others describe what the aircraft is doing around its own axis: pitch (how the nose is moving vertically), roll (how the wings are tipped), and yaw (the sideways motion of the nose). At any one time, there can be a combination of three or four of these: an aircraft in a banked turn is displaying lateral motion and longitudinal motion, but it can also be climbing at the same time. As it turns there is yawing, and as it climbs pitch may change.

In the case of Flight 427, Dellicker was supplying data to fill in some of the blanks, parameters not recorded on the flight data recorder. These included the angle of attack—that is, the angle of the aircraft in relation to the airflow—and the sideslip angle, which is the degree to which the aircraft "skids" in a turn or a roll. Sideslip is easily calculated in a fully instrumented aircraft by measuring the change in air pressure across the nose, but not in a poorly instrumented aircraft like USAir 427. And sideslip was useful to know because the output of a flight data recorder often reveals worrying anomalies—sudden brief changes in altitude or airspeed, for example—that can sometimes be explained by sideslip. Airspeed is clocked by measuring the rate of airflow through a narrow opening, the pitot tube, protruding forward from the fuselage. Altitude is measured by readings of atmospheric pressure, which gets lower the higher you fly. Indeed, the instrument that measures altitude is a type of barometer. Data spikes could also be caused by factors other than sideslip. The wake of another aircraft produces a vortex with lower pressure at its core. Airspeed and altitude data from an aircraft passing through vortices may be momentarily compromised as a result. But in the final analysis, knowing that one thing happened (or did not) enables kinematics to rule something else in or out.

When Dellicker first ran his kinematic reconstruction of Flight 427's final moments two weeks after the accident, he assumed, for the purpose of kick-starting his calculation, that there was no sideslip and gave it a zero value. Predictably, this produced speeds and altitudes far removed from what the flight data recorder showed. Assuming zero sideslip gave the picture of an airplane hitting the ground 100 knots faster than Flight 427 actually did, and 4 seconds too early. Dellicker went back to simulator data for a ballpark figure of sideslip, and this time he got closer, within 10 knots of the flight data recorder airspeed, and the time of impact was close to the recorded time.

Dellicker's kinematic reconstruction suggested that full left rudder had been applied and held at the blowdown limit, or the point at which the pressure of the airflow over the tail of the airplane prevented the rudder from turning any further. Then, as the aircraft rolled over on its back, the pilot turned the wheel fully to the right, in the opposite direction of the rudder, using the ailerons to attempt to level the plane. But there were gaps in this picture. Dellicker still had not produced something definitive to reflect the likely yawing that moved the nose sideways in response to the rudder movement and the rolling that actually turned the aircraft over.

He was missing something else. At the point where the aircraft rolled over, according to the flight data recorder, it was also approaching a stall. The moment at which the wings started to lose lift was indicated on the cockpit voice recorder because the stickshaker, the device that causes the control column to vibrate strongly and noisily in advance of a stall, went off. Because of the way the aircraft was rolling, Dellicker assumed that one wing may have been stalling before the other, which might have increased the yawing motion.

After spending almost 700 hours poring over the data, Dellicker was all but convinced that only the rudder could have produced the amount of yaw that was present. There was the matter of the number one slat, the wreckage of which suggested it might have started to come off the airplane before impact. The slats move forward on the wing's leading edge as the flaps are extended at the rear, and help maintain lift as the airflow decreases when the aircraft slows down on its approach to the airport. The number one slat has a substantial surface area, and there was speculation about what might happen if somehow it poked up perpendicular to the airflow. But this had never happened in the recorded history of the 737, and Boeing engineers thought it more likely to snap off completely if damaged. Dellicker tried factoring it in, but the effect failed to measure up to what was on the flight data recorder. He tried the same thing with the possibility of a reverse thrust deployment, the use of the engine to slow the airplane on the runway after landing. The rear of the engine cowling flips open and forms two sections, called clamshells, that divert the flow of exhaust gases forward and help brake the airplane. They are not meant to come open during flight, but on one tragic occasion they did on a Lauda Air 767 at cruising speed over Thailand on May 26, 1991. All 223 people aboard died as the aircraft broke up in midair as a result. Reverse thrust components found in Flight 427's wreckage indicated it had not been deployed, at least at the moment of impact. But could it have operated momentarily, enough to provoke a fatal dive? Derricker calculated that the kinematic data did not support the possibility of premature reverse thrust deployment.

apter 11   THE FIRST PUBLIC HEARING

At 3:57 P.M. on October 31, 1994, only 53 days after the Flight 427 disaster, another US airliner—American Eagle Airlines Flight 4184—fell from the sky. The plane crashed upside down into farmland near Roselawn, Indiana. As was the case with Flight 427, the pilots of American Eagle 4184 made no coherent emergency call to alert controllers to their plight. Again, investigators were furiously scratching their heads immediately after the crash. Another perfectly good airplane, this time a French-built ATR-72 turboprop less than two years old, had fallen without warning. Sixty-four passengers plus the two pilots and two flight attendants died instantly.

Two days later, after the ATR's flight data recorder arrived at NTSB headquarters in Washington, investigators were plowing through more data than they could immediately deal with. A week after the crash, NTSB Chairman Jim Hall was able to write a lengthy letter to David Hinson, his opposite number at the Federal Aviation Administration, telling him it was likely the pilots had lost control because of a buildup of ice on the wings. Hall's letter, largely drafted by senior investigators, recommended that ATRs avoid icing conditions until the problem could be solved.

A Clinton appointee to the NTSB, Hall was a lifelong Democrat from Chattanooga who had been in charge of road safety in Tennessee before being brought to the attention of the White House as a safe candidate for a senior political nomination. Unlike his predecessor, Carl Vogt, who like Hall was a lawyer but had also been a navy flier, Hall knew very little about aviation before coming to the NTSB. As a state highway administrator, surface transportation had been his forte. When he had trouble understanding something technical, he was known to say, "I apologize, I'm from Tennessee." But despite being new to aviation, Hall was energetic, ambitious, and resourceful. He liked making waves and was not about to let technicalities get in his way. He quickly came to grasp the issue of the small number of flight parameters recorded by flight data recorders on most US airplanes as compared to the much larger number on European-built aircraft. Flight 585 had 5 recorded parameters, Flight 427 had 11, but the American Eagle ATR in the Roselawn crash had 98.

Investigators probing Roselawn had no need for elaborate kinematic reconstructions or extensive performance sessions in simulators. With just about all the cockpit instruments and flight controls recorded, investigators quickly zeroed in on erratic aileron movements, probably caused by icing, as precipitating the crash. The relatively small effort required to solve the ATR crash made a deep impression on Hall, and his zeal on the parameters issue would grow to Ralph Nader-like proportions.

Hall's first opportunity to weigh in on the issue came in January 1995. Ordinarily, when the NTSB conducts a public hearing into an airplane accident, it is glaringly obvious what the investigators are thinking because of the focus on a few selected witnesses in a narrow field of expertise. At the first public hearing into Flight 427, held over five days in Pittsburgh beginning January 23, 1995, the situation seemed dramatically different. Confusion and bewilderment reigned, reflecting the fact that the NTSB investigation, far from being focused, seemed to be all over the place.

The NTSB had booked the hearing into a conference room at the 35-year-old Pittsburgh Hilton and Towers Hotel, an enormous, featureless slab of a building that dated from Pittsburgh's self-styled Renaissance period, when the city started to lure banks and other service companies in a bid to revitalize its declining smokestack economy. A battery of TV cameras lined the rear of the room, their lenses pointed at the dais where the chairman and his board of inquiry sat. The area allocated to the press had filled up rapidly with reporters from national news agencies and local newspapers, as well as specialized aviation publications. Elsewhere in the room—sitting together and mostly keeping to themselves—was a somber, determined-looking group of men and women. Some were holding up photographs or wearing them around their necks. They were relatives of the victims of Flight 427. None of them had ever attended a meeting like this before, and it showed.

Unlike a trial that might be expected to come to a verdict, this hearing was not designed to reach any conclusions. The people with the final say were the NTSB's five board members, White House appointees, by law drawn from both major US political parties, and with five-year terms. Only two of them were present on this preliminary board of inquiry in Pittsburgh: Chairman Jim Hall and Bob Francis, a former FAA staffer. Only the full board could come to a conclusion about an accident, and then only acting on a draft final report prepared by the staff and submitted for the board's approval. Apart from Hall and Francis, those sitting at the dais were all senior NTSB staffers: William "Bud" Leynor, deputy director of the Office of Aviation Safety; John Clark, chief of the

Vehicle Performance Division; Ron Schleede, chief of the Major Investigations Division; and Michael Marx, chief of the Materials Laboratory Division.

Witnesses, by and large, had an easy time. They were sworn in, but that was where any resemblance to a courtroom procedure ended. Tom Hauteur was examined by his boss, Ron Schleede, and summed up the investigation to that point. He showed a video simulation of 427's final moments, based on the information extracted from the flight data recorder. Bill Jackson, the USAir captain who rode in the jump seat of Germano and Emmett's cockpit, testified about the aircraft's next-to-last flight. He was cross-examined by the Air Line Pilots Association's representative, Herb LeGrow, and also by Captain Gene Sharpe, flight safety director for USAir, both of whom sought and received reassurance that the pilots behaved professionally at all times. Michael Marx, the NTSB's lab chief, asked a curious question almost out of the blue: "Where do you put your feet when you are flying the plane?" "On the floor," a bemused Captain Jackson replied.

Certain witnesses who testified also questioned others. So it was that FBI agent William Perry was asked first by Hauteur to describe his investigation, the background to the drug case in which the informant Paul Olsen, killed on Flight 427, was testifying, and the FBI's fruitless search for evidence of a bomb. Then Jack Wurzel, of the International Association of Machinists and Aerospace Workers, wanted to know if there were explosives that left no residue. Perry said he did not know of any. And then John Purvis, Boeing's director of flight safety, asked Perry if he was aware of lightweight explosives that left no traces, and if any of the parts he examined had been first disinfected in a chlorine solution that might remove explosive traces. Perry ducked both of these questions, saying he didn't do the actual inspection. Purvis asked Perry that if only 20 percent of the forward part of the fuselage could be identified, how could the FBI be sure it didn't miss something? We could only examine what was there, was Perry's reply. Ron Schleede put the lid on this line of questioning by asking if anyone had claimed responsibility for bombing the aircraft. Nobody had, Perry replied.

George Green, vortex project engineer for NASA Langley, could be called the hearing's first star witness. In the 1970s he helped put a miniature weather station on Mars, and for the last 15 years had been directing research into the effect of wake turbulence on aircraft. This was of interest because some years earlier Green had done wake vortex tests with a Boeing 727, the same type of aircraft that had been flying just 4 miles ahead of Flight 427 as Delta Flight 1083. Could its wake turbulence have tipped Flight 427 into an uncontrollable spin? Green was assisting NASA with some studies on the wake turbulence generated by the wingtips of new versions of the Boeing 757 and 767, but he also performed numerous tests on the 727 and used the results as a control, to establish baseline data. If anybody had an up-to-date understanding of Boeing 727 wake vortices, it was Green.

Green showed slides of visible vortices caused by test aircraft flying through plumes of smoke flowing from smoke generators atop pylons. The slides showed what looked for all the world like a horizontal tornado. The audience in the temporarily darkened room sat up at this dramatic representation, although their attention faded when Green converted the phenomenon to mathematics: the strength of the vortex was defined by the weight of the aircraft divided by the air density, the aircraft's forward speed, its wing span, and a host of other aerodynamic variables.

As vortices spin off the wing, they have an initial downward momentum that propels them toward the ground, Green explained, until they reach a point at which they run out of momentum and sinking ceases. Then they start to decay, but the rate of decay depends largely on the weather. Windy weather breaks them up, more rapidly nearer the ground amid the turbulence generated by hills, buildings, and trees. Even on warm, still days, convective phenomena, which cause hot air to bubble up from near the ground, are often sufficient to break them up. The wind will blow them to one side or another so they rarely remain directly aligned along the aircraft's track. There are exceptions. If a gentle wind is blowing in the opposite direction of the track of the aircraft, they will remain aligned. And they sink more slowly toward the evening of a warm day, when the air is cooling and the ground is slowly giving off the heat it accumulated earlier.

It was on this sort of evening that Flight 427 crashed. Green had done the math on the weather that day and how it might have affected the vortices from Flight 1083. He reckoned there was a strong probability the vortices would have descended about 300 feet by the time they reached Flight 427's intended flight path. By coincidence, Flight 427 was flying about 300 feet lower than Flight 1083 and about 70 seconds behind it. It was, said Green, one of the classic things the *Airman's Manual*, the bible of flying, warns against: "You don't want to be below and behind another airplane."

"Did Flight 427 hit the vortices of Delta Flight 1083?" Tom Jacky asked Green directly. "The sky is a big place and it's impossible to prove any airplane went through a very small region, but it was certainly in the right place to be hit," Green told the room, once more hushed with anticipation. Green, who had been involved in helping Boeing program wake vortices into its simulator, admitted that he and Boeing had disagreed on the size of the core of the vortices Flight 427 might have encountered. Yet, even at the higher core strength, the effect on the aircraft would have been relatively minor. Green imagined it would have rolled between 10° and 30° and could have recovered relatively easily. If a 30° roll were the worst one could expect from the vortex of a 727, what would it take to invert a 737? Michael Marx wanted to know. "I don't think there is anything out there that can generate a strong enough vortex to flip a 737 upside down," Green replied. "Not even a 747?" asked Marx. "It wouldn't flip it upside down," Green insisted.

The next witness was Jim Kerrigan, Boeing's principal engineer for aerodynamic stability and control on 737, 727, and 707 aircraft. He had directed the rotor simulations for Flight 585 and had them adapted to produce wake vortex simulations for Flight 427. He reported that pilots who had encountered the wake vortices in the Boeing simulator said they were similar to what they would expect in real life. As far as Flight 427 was concerned, Kerrigan told Tom Jacky, a wake vortex encounter could have initiated the incident, but could not have been responsible for its continuation because the aircraft would have been in the vortex for just 5 seconds, according to his estimates. However, he challenged Green's assertion that a 30° roll would have been the maximum disruption a vortex could cause. To prove his point, he screened a video of one run in the simulator in which the aircraft was flown into a vortex on autopilot. It rolled some 50°, although the autopilot was able to recover level flight and get back on course without any pilot intervention.

There were hopes that FAA engineer Michael Zielinski would be another star witness at the hearing. Normally, Zielinski helped the FAA evaluate new or modified aircraft designs, but since October 1994 he had been in charge of a special probe into how the Boeing 737 was put together. There had been anticipation that Zielinski's investigation, called a Critical Design Review, would be completed in time for discussion at the Pittsburgh hearing. Zielinski appeared at the hearing and had much to contribute, but readily admitted that his design review was unlikely to be published before May 1995. Zielinski's probe arose out of a concern at the highest levels of the FAA that the airplane that crashed at Aliquippa might have contained flaws that a more modern design would have eliminated. New aircraft designs must pass the FAA's rigorous certification process before entering service, and the rules had been considerably tightened since the 737 was first approved in 1967. Yet the process in place to evaluate the 737-300 differed little from that which governed the certification of the original 737-100 and the 737-200, the model that crashed at Colorado Springs in 1991. By and large, the 737-300, which first flew in 1984, was certified according to a late-1960s set of regulations. Had it been a brand-new design, regulators would have taken a tougher stance and applied stricter 1980s regulations. But because it was a variant of an earlier, already-sanctioned design, regulators were obliged to operate on an "if-it-ain't-broke-don't-fix-it" basis. Only those parts of the airplane that were radically different, such as the engines, would be subjected to a more modern assessment. The remainder would be judged according to a 25-year-old rulebook. "Grandfathering" was the term used by the industry to describe the process.

Was there anything about the 737-300 that they might not have approved if it had been a brand-new design? That was the question Don Riggin, manager of the FAA aircraft certification office in Seattle, asked his boss, Tom McSweeney, on the phone a few weeks after Flight 427 had crashed. Both men were puzzled that no clues were emerging as to the cause of the accident. At the back of their minds was the continued unsolved mystery of Flight 585. Both were aware of eerie similarities in the final flight paths of the airplanes.

When the 737 was first certified in 1967, hazard assessment was done on a probability basis. In other words, if there was an extremely remote probability of something going wrong, it could be ignored. A decade later, that approach was changed, but only for completely new aircraft designs. Models that were derivatives of older aircraft could continue to claim Grandfather rights and be assessed under the older regulations. An advisory circular numbered 251309 had been issued by the FAA in the 1970s, outlining a tougher approach for evaluating aircraft designs predicated on the *possibility* of something happening, not just its probability.

With Flights 585 and 427 in mind, Riggin appointed Zielinski to head a Critical Design Review of the 737 and instructed him to ask himself if he would approve this airplane by the newer rules. "The decision was made that the effort could be without any inhibitions, inhibitions from the standpoint of the probability of the occurrence. It was more of a hazard assessment, a qualitative hazard assessment," Zielinski explained to NTSB investigator Greg Phillips at the hearing. "We were free to challenge the applicability of our own regulations." Focusing on the lateral and directional flight controls—ailerons, spoilers, and rudder in addition to various slats and flaps—Zielinski and his team paid particular attention to what might happen to a 737 traveling at 190 knots at Flaps 1, the configuration adopted by Flight 427 shortly before it crashed. When the 737 was being originally certified, the failure of a flight control could be ignored if it could be shown to be extremely improbable. Now Zielinski's team was asking if the aircraft could be landed safely using alternative means should any flight control fail, regardless of probability.

The FAA team based itself at Boeing's Renton plant near Seattle and spent some time in the Boeing simulator working through a test plan containing some 50 flight scenarios. Boeing was cooperative. "They were attentive to the point of not arguing that a line of inquiry was so improbable as to be not worth worrying about," Zielinski told Ron Schleede at the hearing. "Boeing even refrained from arguing the probability of failure, so we were discussing failures irrespective of their probability." Zielinski's broad objective was to track down any failures that could cause a jam or an uncommanded deflection of any control surface. Latent failures, mishaps that could go undetected but were within the bounds of possibility, would also be identified.

The simulator mission had two parts. First, putting the simulator through its paces to see how it coped with all the failures, real and imagined, that they could throw at it. And second, sifting through the archives at Boeing, the FAA, the NTSB, and NASA to come up with an historic profile of the 737's failings and decide if solutions were sufficient. Existing methods of dealing with flight control failures were considered to see if they were realistic. Or did they increase the physical and mental workload of the crew to the extent it might be incapable of dealing with any further emergencies? And did those methods give the crew enough time to identify and react to a failure before an airplane was totally out of control?

In addition to the complicated matter of flight controls, the design review team had problems with certain other aspects of the 737. They were particularly alarmed at the lack of protection for the aircraft's hydraulic system reservoirs, all three of which were stored in the wheel well of the main undercarriage. At the time of its certification, wheel-well components were assumed to be at risk only from a high-pressure blast of nitrogen gas from a bursting tire. No consideration had been given to pieces of tire tread or debris from a disintegrating wheel being flung into the wheel well. Such events had never occurred and therefore were thought to be extremely unlikely. Indeed, Boeing later received approval to remove protective screens from the wheel wells on the grounds that they were no longer necessary.

After a team visit to the Parker Hannifin plant in Irvine, California, there was concern with the possible vulnerability of the rudder power control unit, which directs the extent and rapidity of rudder movement, if chips of stray metal found their way into its servo valve. Although the valve was designed to shear off any metal that jammed its orifices, the team wondered if it always generated enough force to accomplish this. And, while impressed with bench testing of units being serviced, they noted that there was no way to check that the unit actually worked properly when put back on the airplane.

Neither was the team happy after a review of the 737's yaw damper, the quasi-autopilot that keeps an airplane flying straight, despite the tendency of all swept wing jets to shimmy in a phenomenon known as Dutch roll. When it detects an uncommanded deviation from straight flight, the yaw damper commands small rudder movements to bring the plane back on course. In one 12-

month period alone, some 200 yaw dampers had been classified as failures for one reason or another. Of those, 130 were caused by damaged bearings in the rate gyro, the sensitive component that detected any yaw. This was mostly blamed on increased vibration from the newer, more powerful engines aboard the 300 series. Another 28 failed because of more normal wear and tear, including short circuits. But no reason could be found for the failure of the remaining 42 units. The team was also alarmed to discover that Honeywell/Sperry, which manufactured the 737 yaw damper, was unaware of one short-circuit condition that could cause a 3° rudder deflection for up to 2 minutes at a time. However, the company was aware of other circuitry defects that could result in yaw damper-induced rudder deflections, although usually of shorter duration.

In mid-November 1994, the Critical Design Review team flew to Seattle for a session in the Boeing M-Cab simulator. There were some more surprises in store. For example, switching off the autopilot in the middle of a large aileron movement and taking the pilot's hands off the wheel resulted in rapid, large rolling movements, up to 44° per second at higher speed. If the pilot failed to grab the wheel quickly enough, an unnerving bank angle of 60° or more would result. They also found that loss of "feel" in both rudder and aileron controls often resulted in excessive inputs by the pilots. Jamming the ailerons at up to 20° could be overcome by using the spoilers on the wings, panels that can also function as ailerons in an emergency. The spoilers are controlled by wheel commands through a transfer mechanism, but the CDR team concluded "Long term flight to a successful landing was questionable, due to pilot effort required, and the onset of pilot fatigue." In other words, the aircraft was controllable but they doubted if many pilots had the strength and stamina to keep it flying to a safe landing.

The Critical Design Review team also simulated a rudder hardover, in which the rudder is deflected and jammed to one side as far as it will go, which depends on the airspeed: the slower the airplane is flying, the further the rudder can travel. By and large, they found that the airplane was controllable using the ailerons, bearing out Boeing's contention that they offered pilots enough control to counteract the rudder. Then they tried the same thing at 190 knots airspeed at Flaps 1, the configuration of Flight 427 before the pilots lost control. The airplane was controllable but recovery was slow and required precise control of pitch and airspeed by the pilot. If the pilot failed to react quickly enough at higher airspeeds and altitude, it was not long before the aircraft was flying upside down.

The team warned that the problem of the standby rudder actuator, identified in the wake of the crash of Flight 585 in Colorado Springs and the subject of repeated inspections since then, had still not been solved. Furthermore, there were insufficient means of detecting faults in the main rudder power control unit, and the intervals between maintenance checks on it and other control surface mechanisms were "excessive." Shortly before Christmas 1994, a Critical Design Review delegation went to McDonnell Douglas's Long Beach, California, plant to see how the other half of the US civil airplane manufacturing sector did things. They were impressed with the ability of McDonnell Douglas's more powerful rudder power control units to handle stray metal chips; the weaker Boeing servo valves could be disabled by tough pieces of debris. The team also liked the use of split rudders with dual systems to handle jamming, and noted with approval the use on some McDonnell Douglas airplanes of an easy hydraulic shutdown of a jammed rudder, which could then be controlled by a direct cable system. And regaining control of a McDonnell Douglas airplane after a rudder jam required less physical force than its Boeing 737 equivalent. McDonnell Douglas also insisted on tighter control over hydraulic fluid contamination than Boeing: while McDonnell Douglas specified routine inspection and replacement if its contamination levels were exceeded, Boeing said fluid contamination was an issue for the airlines, not for them.

Later in the hearing Jim Hall asserted himself, to the acute embarrassment of Boeing's representatives, by repeatedly reminding them that flight data recorders aboard some of their airplanes covered the least number of flight parameters of all major plane makers. Even Boeing's kinematic expert, Harry Dellicker, received short shrift when he gave evidence. Hall dryly remarked in his slow drawl, "I only wish you had more parameters to work with. In this accident we obviously don't."

When he testified, Dave King, chief investigator for the UK's Air Accidents Investigation Branch, was asked how many parameters British Airways recorded on their aircraft and answered that the total ran to hundreds. Hall was incredulous. "And they do that voluntarily? Their government doesn't have to come in and tell them to do that? They're doing it on their own?" he asked King. Recalling that British Airways was a shareholder in USAir at the time, Hall added his hope the British would exert some influence in connection with flight data recording. Later the same day he nodded approvingly as senior NTSB investigator John Clark closely questioned Boeing's chief engineer, Jean McGrew, about his company's minimalist approach to flight data recorders. Clark pointed out that Boeing's main US rival, McDonnell Douglas, routinely shipped aircraft with far more recorder parameters than Boeing, irrespective of what rule applied to them.

Later Hall added his own opinion. "I think it's very unsettling to the flying public in the United States that there are airplanes going out of the Boeing factory overseas to Europe that have more sophisticated flight data recorders, recorders that would have possibly provided the information that would have solved not only this accident, but possibly the one in Colorado Springs." Hall delivered this little speech in his usual folksy manner, but its meaning was lost on no one at the hearing. John Purvis, director of air safety investigation at Boeing, squirmed silently in the witness box while Hall read aloud a letter he was drafting, asking for comments by parties to the investigation on what additional parameters they thought were necessary to be retrofitted to all US airplanes. It was clear to his audience what Hall meant by US airplanes: those manufactured by Boeing.

A month after the Pittsburgh hearing, on February 22, 1995, Hall issued his first Safety Recommendation arising out of Flight 427 to the FAA. It cited the situation concerning flight data recorders on Flights 427 and 585, in contrast to the Roselawn ATR, and called for improved flight data recorders to be fitted to all US aircraft. As Hall later remarked, "One of the greatest enemies of safety is the grandfather clause where older systems get grandfathered in, and the public has no way of knowing which of these systems they are flying on and what type of technology they have." As examples, he cited 12 other incidents investigated by the NTSB over the previous few months in which 737 aircraft had experienced uncommanded movements, then went on to say this:

> ... Like 79 percent of all US-registered Boeing 737s, the airplanes involved in the incidents were manufactured prior to May 26, 1989; consequently, they were required by current regulations to record only the five basic FDR parameters. As a result, critical objective data were not available from the FDRs, and investigators had little more than the flight crews' subjective recollections of these dynamic events...."

Contrast that, he continued, with the European situation. British Airways quickly solved the mystery of a Boeing 747 that experienced a sudden nose-down pitching movement on taking off from Heathrow in 1993, thanks to the hundreds of parameters recorded on the aircraft's quick-access recorder. Investigators from the UK Air Accidents Investigation Branch quickly discovered that there had been an uncommanded movement of the right elevator on the tailplane just as the landing gear was being retracted. A surge of hydraulic pressure was blamed and the AAIB called for changes to the 747's hydraulic system to prevent a recurrence. Similarly, in 1993, Air France had 206 flight data parameters recorded per aircraft, which helped it get to the bottom of three uncommanded rudder deflections on Boeing 737s. The Air France aircraft were of the same vintage as USAir Flight 427, yet they employed flight data recorders nearly 20 times as powerful.

Hall knew better than to ask for the moon from cost-conscious US airlines, so he recommended that sensors and recorders tracking only 21 parameters, including the current 5, be made mandatory for retrofitting to older aircraft. He also agreed that, apart from Boeing 737s, any older aircraft not likely to meet new noise suppression regulations and likely to be scrapped or retired before the end of 1999 could be exempted. Nevertheless, he invited airlines and manufacturers to voluntarily fit equipment that would record even more parameters, as was routine in Europe. Acutely aware that the airlines and plane makers would object on grounds of cost, Hall had some back-of-envelope calculations ready to refute their inevitable objections. While some 1,000 affected Boeing 737s could be modified for $20,000 each, it might cost more than three times as much to retrofit some larger airplanes. But translating the estimate into cost per passenger flown for the remaining service life of the airplane amounted to just 7 cents. "The safety board believes that public safety outweighs the seven cents per passenger cost of equipping older airplanes to record more FDR parameters," he wrote to FAA administrator David Hinson.

In the wake of the hearing on Flight 427 at Pittsburgh, Michael Zielinski's Critical Design Review team wound up its business. It found no major faults with the 737, nothing that might have caused the crash of either Flight 585 or Flight 427, and nothing to warrant immediate regulatory action. Nonetheless, it made 27 recommendations to the FAA. Most arose out of the matters aired during the hearing, but there were additional scenarios they wanted addressed. One of these, not addressed in the original certification of the 737, was the issue of rudder cables being severed in the event of an engine exploding catastrophically. In that event, an airplane could become totally uncontrollable due to the asymmetric thrust of the remaining engine. Giving crews logical and practical responses to various kinds of failures was another issue the team wanted raised, as was the lack of a service life specification for some control cables.

Some of the strongest recommendations centered on the 737 rudder and its components. The review team was particularly concerned about the relationship of the rudder and the yaw damper, and warned that any yaw damper malfunction that caused the main rudder power control unit servo valve to be held open could result in a rudder hardover. "The CDR team believes that all the failure modes of this mechanism have not been fully examined,"

Zielinski's report stated. Yaw damper failures aboard 737s had to be brought down to an acceptable level, and maintenance techniques for the power control units, both rudder and aileron, had to be improved.

Perhaps the most significant of Zielinski's recommendations was number 27, which called on the NTSB to reopen the investigation into Flight 585 and combine it with the Flight 427 probe. "Through the critical design review effort, the team took a fresh look at the B737 flight control design and certification and believes there is merit in taking a similar fresh look at all of the data gathered on both accidents. Combining a fresh look at the accident along with the data learned from the CDR could shed new light on the cause of these accidents," Zielinski concluded.

But his review did not have anything like the impact he hoped it would have among the members of the Systems Group, the NTSB's investigation into Flight 427. Yes, the NTSB people all agreed it was a useful report, but it told them little they did not already know. As far as combining the Flights 585 and 427 investigations, that had already happened on an informal basis; it was only a matter of time before investigators officially linked the two probes. There was some disappointment that Zielinski's team had not devised tests to further probe the reliability of the control surface systems they were investigating. And there was a lingering suspicion that Zielinski's team was not as independent as the NTSB people would have liked it to be. "The CDR was the result of a compromise agreement within FAA," said John Cox, the ALPA representative on the Systems Group investigating Flight 427. "We knew this. There were several other drafts before the final 'product' was released. We were aware of the drafts and were even told of some of the items. Over time the CDR mellowed. The initial drafts were more critical of the 737. The CDR's release did not change our opinion. We knew where the focus should be. There was a problem in the tail. We had to find it."

hapter 12 HUMAN FACTORS

At a conference of all the investigating groups on January 19, 1995, just a week before the first public hearing on Flight 427 in Pittsburgh, Tom Hauteur appealed to all members to seriously consider any other issues they felt should be investigated. After the hearing, on February 9, the Air Line Pilots Association (ALPA) came forward with the suggestion of reconvening the cockpit voice recorder groups from both Flight 427 and Flight 585, and comparing tapes to see if they held any unusual sounds in common.

In mid-February, Boeing's director of flight safety, John Purvis, wrote to Tom Hauteur and suggested that a full-fledged Human Factors Group be named. Boeing had already withdrawn its delegate to Chuck Leonard's Operations Group as an unsubtle hint that its investigation was unlikely to uncover the cause of the Flight 427 crash. Instead, Boeing had wanted to probe the cockpit voice recorder and the wake turbulence theory. Purvis wanted the NTSB to consider the possibility that the pilots had overreacted by incorrectly using the rudder or other flight controls when the plane was knocked about by the wake of the aircraft preceding it into Pittsburgh. To put it bluntly, by "human factors" Purvis and Boeing meant pilot error.

When wake turbulence was discussed at the January public hearing, an FAA test pilot, Les Berven, had described an unexpected run-in with the phenomenon as "nothing like you encounter in normal flight ... it basically feels like some giant hand grabbed the airplane and just took it right away from you...." In his letter to Hauteur, Purvis quoted the test pilot's comments, then added: "The sounds experienced and the comments made by the flight crew during the first five seconds of the encounter with wake turbulence corroborate Berven's testimony." Nevertheless, Purvis's letter seemed innocuous compared to a 5-page attachment that made the pilots involved in the investigation see red.

Purvis referred to separate UK Royal Air Force and FAA studies on pilots' sometimes improper reactions to unexpected events, and stated that the NTSB should explore "whether the Flight 427 flight crew could have responded to the unexpected and startling encounter with significant wake turbulence by

(1) making an inadvertent application of left rudder, or (2) having an accidental or cognitive failure that led to an application of left rudder." A related question, added Purvis, was whether the crew's training and medical records showed anything about their ability to diagnose emergencies, handle stress, and use the flight controls correctly. Citing five examples of incorrect rudder commands causing accidents or emergencies, Purvis asked if the same thing could have happened on Flight 427. Was there any record of the crew misapplying flight controls? Pilot movements of the control column were the only flight controls preserved on the flight data recorder, and the data showed that the crew had pulled back hard on the "stick" as far as possible once the autopilot had been disconnected. Was this appropriate? What role did the control column commands have on the crew's ability to recover the aircraft? And if they made inappropriate control column commands, was it also likely they responded with inappropriate use of the rudder? How did the crew interact in the emergency? Were both pilots trying to fly the aircraft when only one pilot should have had his hands and feet on the controls at any given time?

USAir objected vehemently to Purvis's questions, which they took more as suggestions of blame rather than dispassionate queries. "We believe human factors inquiries are inappropriate and ill-advised," wrote USAir's accident coordinator, Gene Sharp, to NTSB chairman Jim Hall. Urging him to drop the idea, Sharp added: "There is no credible evidence that human factors played any role whatsoever in this [crash]. We believe that, by engaging in this new line of inquiry, the NTSB will lend credibility to the spurious suggestion that the pilots are somehow at fault in this accident."

ALPA was even less pleased. When the organization learned of Purvis's letter, the reaction at its Herndon, Virginia, headquarters was furious. "Some of the Boeing contentions are unsupported and erroneous," ALPA's accident investigation coordinator Herb LeGrow retorted in a heated response to Hauteur that was copied to Jim Hall and several other parties to the investigation. "In the Boeing submittal, there is an inference that the flight crew caused the accident involving USAir 427. There are no facts to support such a conclusion. This document attempts to use unrelated accidents and incidents to create doubt about the human performance of USAir 427's flight crew." LeGrow was also critical of Purvis's use of Les Berven's testimony at the hearing and his linking it to the transcript of Flight 427's cockpit voice recorder. "The attempt to relate Mr. Berven's testimony to the CVR transcript without a complete understanding for the loss of control is unfounded.... This document presumes to tell the NTSB what questions they should ask regarding flight crew performance. This list of questions is ambiguous and, at best, poorly phrased. In several cases the question has already been answered by the physical evidence."

LeGrow then refuted Purvis's points one by one. To the suggestion that the pilots might have used the rudder to control rolling, LeGrow replied: "The question is so basic that it is meaningless. All aircraft can use rudder to offset roll if the rudder is operating properly. Boeing well knows this." LeGrow also rejected the suggestions that comparisons might be drawn between Flight 427 and the list of incidents and accidents cited by Purvis where crews had made inappropriate rudder commands. "A review of these incidents shows there is no connection with USAir 427."

LeGrow was scathing about Purvis's suggestion that inappropriate use of the control column could also imply inappropriate rudder use. "This is appalling," he wrote. "There is no evidence that there was improper rudder input by the crew. In spite of this lack of evidence, this document used conjecture and speculation to attempt to say that there was both improper rudder and control column usage. This unsupported speculation has no place in accident investigation."

For the first time in the investigation, the battle lines were emerging clearly. LeGrow made no secret of the fact that ALPA was now nailing its colors to the mast by declaring its belief that Flight 427 had been downed by a hardware problem. LeGrow wrote:

> ... All evidence collected to date in this investigation point[s] to the strong possibility of a mechanical malfunction in the tail of the aircraft.... Everyone recognizes the manufacturer's product liability problem. The issues of civil litigation should not be allowed to infiltrate an NTSB investigation. The traveling public deserves the answer to what truly caused this accident. In parallel with the Human Factors Group's activities, we must continue to pursue the aircraft systems, and structures, for clues that would enable us to identify the cause of this accident. By focusing on the human factors aspect of this investigation, we do a great disservice to the traveling public and lay the foundation for another horrible accident to occur before we determine what happened to USAir 427....

The language repudiating Boeing's position could hardly have been stronger.

Despite the uproar, Malcolm Brenner, following Hauteur's directions, proceeded to form his new Human Factors Group by requesting nominations from the various parties. ALPA appointed Robert Sumwalt, a USAir 737 pilot. Boeing nominated two people, Curtis Graeber, an ex-NASA psychologist with a Ph.D. and extensive experience in the human factors field, and Captain Michael Carriker, a Boeing 737 test pilot. Carriker was a 12-year veteran of the US Navy and had graduated from test pilot school and become an instructor there. He also had a spell as a test pilot instructor in Britain. The FAA nominated Phyllis Kayten, a former NTSB human factors investigator with a Ph.D. in psychology who now worked as a human factors specialist in the FAA's liaison office at NASA's Ames Research Center in California.

Malcolm Brenner's involvement with aviation accident investigations was totally unintended. After obtaining his Ph.D. from the University of Michigan in 1976 and publishing his dissertation on voice analysis as a means of measuring stress, he was hunting for a lecturing job as a stepping stone to what he hoped would be a faculty position at a college or university psychology department. Out of the blue, he was approached by a law firm representing an airline that needed an expert witness to help its defense in a lawsuit following the crash of a DC-10 in Mexico. Someone had read his dissertation and thought he might be a useful expert witness. Brenner donned his only suit and set off to a consultation where he was given a half hour of cockpit voice recording to analyze. His appearance in that case led to others, and before long Brenner became a sort of freelance star witness for airlines (and sometimes passengers or their surviving relatives), helping lawyers build cases or refute others on the strength of his analysis of cockpit tapes.

The NTSB was looking for a suitably qualified person to fill a new position—Human Factors investigator. Brenner, with his psychology background and his experience in analyzing cockpit voice recordings, seemed to fit the bill. In addition, he had learned to fly and earned his pilot's license, and in the eyes of the NTSB, this gave him all the more credibility in his analysis of cockpit tape recordings. Brenner, too, thought the fit between himself and this new job was a good one, especially since he'd recently been tiring of the endless round of court appearances and the stress of aggressive questioning. So he accepted the NTSB's offer gratefully.

Interestingly enough, Brenner first achieved distinction as an NTSB investigator not in a plane crash probe but in one of the most notorious marine disasters of the twentieth century. On March 24, 1989, a gigantic oil tanker, the *Exxon Valdez,* ran aground in Prince William Sound on the south coast of Alaska, splitting open and creating the greatest oil spill in US history.

A glacier continuously drops icebergs into Prince William Sound, and the supertanker's captain, Joseph Hazelwood, decided to depart from the main shipping channel to avoid an iceberg that was drifting across his course. He unwisely set the ship on autopilot and then left the already tired and overworked third mate in charge while he retired to his cabin. The *Exxon Valdez* grounded on Bligh Reef and released 11 million gallons of North Slope crude oil, creating an environmental disaster of epic proportions.

Hazelwood had two drunk-driving convictions and a history of being admitted to hospitals for treatment of alcohol abuse. Witnesses reported he had been drinking on shore the afternoon before the accident, though when he boarded his ship he did not appear impaired. A blood test taken almost eleven hours after the accident revealed a blood alcohol content of 0.06 percent. Worked backward by using the average blood alcohol metabolism rate, that level might suggest a blood alcohol content of about 0.2 percent about the time of the accident, a dose strong enough to intoxicate most people.

Malcolm Brenner was put in charge of the human performance investigation for the Office of Marine Safety of the NTSB in the *Exxon Valdez* accident. The question Brenner needed to answer was whether Hazelwood was intoxicated to the point of exercising impaired judgment. Brenner decided to try something never attempted before in a US investigation. He wanted to recreate a subject's physiological condition from voice recordings. As a graduate student, he had researched the effect of stress on speech in a NASA-funded study. He wondered whether a similar analysis could be done to measure intoxication.

There was no maritime equivalent of a cockpit voice recorder, but the *Exxon Valdez* did have the equivalent of an air traffic controller in the shape of the US Coast Guard Traffic Center for Valdez Bay. Like the FAA, which routinely records all air traffic control radio transmissions, the Coast Guard records all its communications with shipping personnel. As a result, Brenner found himself in possession of recordings of radio traffic between the Coast Guard watchkeepers on duty and the *Exxon Valdez's* bridge over a three-day period. Hazelwood's voice is heard on the tape 33 hours prior to the accident, an hour beforehand, immediately afterward, an hour afterward, and finally nine hours afterward.

Brenner sent copies of the tapes to Mark and Linda Sobell, both professors in the Department of Behavioral Science and Psychology at the University of Toronto. The Sobells specialized in alcohol addiction and its effect on behavior, and in the 1970s, they performed groundbreaking work on the effect of alcohol on speech, comparing the way people spoke when sober and when drunk. The contrast between Hazelwood's speech when sober and when apparently intoxicated was so strong that the Sobells wrote in their report that it didn't require anybody with their training and experience to spot the difference.

Brenner next went to the Speech Research Laboratory at Indiana University, where experts subjected the Hazelwood recording to a computerized spectral analysis enabling them to produce graphical representations of individual syllables and consonants. They were able to measure, in milliseconds, the length of time it took Hazelwood to say certain key words, like the name of his ship, that were frequently repeated. Like the Sobells, the Indiana experts found that Hazelwood made common alcohol-impaired speech errors, such as when "s" becomes "sh." This was especially noticeable in his pronunciation of the word "Exxon." Normally he pronounced it "eggson," but about the time of the accident he was saying "eggshawn."

Hazelwood's speaking rate at various times was also revealing. Thirty-three hours prior to running aground, he fluently snapped off the words "Exxon Valdez" in 700 milliseconds (0.7 seconds). As he was reporting the accident with his ship on the rocks and hemorrhaging oil into the ocean, one might have expected his speech to become more rapid with stress and excitement, but the opposite was the case. It took him 1,100 milliseconds (1.1 seconds) to say "Exxon Valdez," 56 percent longer than when he was sober. Nine hours later he was taking less than 900 milliseconds. There was little doubt that Hazelwood was intoxicated to the point of impairment when his ship ran aground, the NTSB concluded, in the process adding Malcolm Brenner's new analysis techniques to the crash investigator's repertoire.

In the period of thaw in the early 1990s between the former USSR and the US, Brenner came across the work of Alfred S. Belan, who was in charge of the speech laboratory of the Interstate Aviation Committee in Moscow. Belan's research was based on some 300 cockpit voice tapes, several times the 50 or so reported in the entire Western academic literature. That claim alone, if it were true, would make him a very interesting exchange scholar indeed. Brenner, by now well known in his field, arranged for Belan to come to Washington for a visit.

Russian interest in the field of speech analysis went back to a study by US Navy researchers analyzing the famous recording of the radio reporter, Herb Morrison, who witnessed the burning of the *Hindenburg* on May 6, 1937, at Lakehurst, New Jersey. The study attempted to relate the changes in his speech to his emotional state as he witnessed the horror of the burning German dirigible and included his utterance of the famous words, "Oh, the humanity ... and all the passengers. . . ."

In the US, this avenue of research became an academic cul-de-sac, but it sparked intense interest among technology-starved Russians, who saw it as a low-tech means of analyzing accidents simply by adopting a much more clinical approach to the subject. Belan developed useful analytical tools that enabled the Russians to delve deeply into the state of mind of pilots experiencing emergencies. Belan offered Brenner an interesting analogy to explain why the Russians were more successful than the Americans in exploiting such techniques. America spent millions developing a pen that could write in the zero gravity of space; Russian cosmonauts used a pencil.

Belan's analysis of speech assumed three levels of stress. The first level of stress is actually positive: it is motivating and it focuses a pilot's attention, which means he performs better. It could be detected by measuring the increase, up to 30 percent, in the rate of a pilot's speech and the increase in loudness, among other factors. The next level of stress is intermediate: pilots are still functioning but are nearing the threshold of panic. They drop unnecessary adjectives, omit checklist items, and operate—perform tasks, or attempt to perform them—more quickly. Their speech becomes more rapid and increases in loudness, and they often repeat themselves, as if to make sure the hearer understands. The third level is sheer panic, when pilots become fixated on just one thing, and not always the right thing. Speech becomes incomplete, syllables are omitted, and words are often unvoiced or swallowed. Word choice becomes poor and grammar improper, and the usual measures of speech rates and loudness show increases of upward of 200 percent.

In addition, Belan advised Brenner, it is not always profitable to listen to *what* the pilots are saying. Examine *how* they are speaking. Sometimes you need to ignore their meaning and even remove the words from the recording in order to focus on the pilots' breathing. Brenner also learned from Belan how to tell if a pilot is suffering from hypoxia, or oxygen starvation, as a result of a breakdown in the aircraft's pressurization system.

As a result of Brenner's work, the NTSB was probably the first safety agency in the world to formally recognize human factors as a discipline in its own right. But the development was not without controversy. A rash of jokes greeted the entry of psychologists into crash investigation: plane crash probes were now "the study of the id by the odd," according to one. And some traditionalist tin-kickers couldn't see why they needed a psychologist to tell them a pilot had screwed up when that was usually as plain as the nose on your face.

Brenner conceded that considering human factors involved matters that formerly were grouped under the broader heading of pilot error, but it was no longer sufficient to simply say "this pilot is not up to standard, so that's a probable cause of the crash." Brenner saw his role as bringing much more to an investigation. Is there some flaw in the overall system that causes pilots to screw up? Or that encourages pilots, or at least makes it easier for them, to make bad choices? Is there some pitfall in regulating flight and duty times that causes pilots to fly when they are fatigued? Is their training adequate? Are there hidden corporate pressures to make more trips or fly faulty aircraft? Sooner or later, tired pilots were going to run into bad weather or have something go wrong early in the morning when they were not at their sharpest because they needed sleep. Brenner sought to get a sense of what a human being might reasonably be expected to do and to determine if too much was being expected of some pilots.

Initially, and perhaps inevitably, Brenner was given jobs like interviewing next of kin after a fatal crash, a job nobody liked doing because it brought him too close to the human tragedy and misery involved in any accident. Brenner also found himself being asked to interview witnesses, review crew medical records, and handle toxicology testing. After Flight 585 crashed in Colorado Springs, he spearheaded the human factors group as a subsidiary of the Operations Group investigation. After Flight 427 crashed, Brenner again headed a sub-group, this time under Chuck Leonard in the Operations Group, as lead after lead was chased down and eliminated.

Robert Sumwalt, the USAir pilot assigned by ALPA to Brenner's Human Factors Group in the Flight 427 investigation, was impressed when he learned that Curt Graeber would be joining the group as one of two Boeing representatives. R. Curtis Graeber was a psychologist who had moved from the US Army to NASA, where he became attached to the NASA Ames Fatigue/Jet Lag Program, an initiative undertaken at the urging of the US Congress. He participated in research on a broad range of crew fatigue issues, from the impact of jet lag on pilots' sleep patterns to the difficulties posed by flying night cargo routes. The issue had a dramatic bearing on flight safety, and the research coming out of NASA sharpened awareness of the need for effective crew rest periods. It gave rise to the "NASA nap" after research discovered that pilots who took short naps on the flight deck (forbidden under FAA regulations) actually performed better. Sumwalt had listened in awe to Graeber's presentations at conferences, often paying out of his own pocket to attend them whenever the opportunity arose. Now he would be working side by side with a giant of aviation psychology in a bid to solve the mystery of Flight 427.

Sumwalt, a union activist since 1987, had first become involved with crash probes as an ALPA representative on a meteorological group—a job he found too technical for his tastes. He had always been interested in the effect of what was once known as human frailty on air accidents. How accidents happened because pilots were inattentive, omitted checklists, failed to warn an erring superior officer, misread instruments, underestimated the severity of weather, or made false assumptions—all this fascinated him. Depending on how you interpret the statistics, human factors can historically be blamed for perhaps 70 percent of all air accidents, most taking place on approach and landing. As human factors developed into a crash investigation discipline in its own right, Sumwalt started collecting research papers and attending conferences on the subject.

By 1991, ALPA had organized its own human factors committee and invited Sumwalt aboard. He reported on conferences for aviation magazines, and ALPA representatives on crash investigation teams sent him draft reports to evaluate. Then NASA made him a consultant for its Aviation Safety Reporting System, a program that allowed pilots to anonymously highlight errors and reveal the circumstances of near-accidents without precipitating a full-scale NTSB or FAA inquiry.

Flight 427 was Sumwalt's first full-scale fatal air crash as a human factors investigator. Two months earlier his family had a close encounter with air disaster when a USAir jet crashed near his hometown of Charlotte, North Carolina, following a wind-shear encounter. His wife's brother-in-law was a passenger and was among the injured. On being discharged from the hospital, he described the crash, in which 37 people died, as "an act of God." Sumwalt wasn't so sure. He was interested less in the possible involvement of the Almighty and more in learning why the pilot had decided to fly into a thunderstorm positioned across his final approach.

Sumwalt was assigned to the original Flight 427 Human Factors Group on the day after the crash, September 8, 1994. Like other group leaders, Malcolm Brenner was preparing his members to suit up in biohazard gear prior to visiting the site when Tom Hauteur stopped by and advised them not to go. "This is a very bad scene and I don't think you guys need to go there," he said. "If you don't need to go there, I don't want you there." Sumwalt was relieved. Just a few months earlier he had been part of an ALPA group that established a counseling service for members working on crash investigations after it was shown that the mere act of working on a serious accident could have a devastating effect on a pilot's personal life.

The original 427 Human Factors Group was a subgroup of the Operations Group run by Chuck Leonard. Sumwalt got sucked into the mundane, but essential, chores of tracking down witnesses and getting their statements. He went to Chicago and interviewed the mechanics who had seen the doomed aircraft off, and the pilot passenger in the cockpit, Captain Bill Jackson, who had inadvertently tripped the jump seat microphone. He even visited the hotel the pilots had stayed in the night before the crash to see if they had ordered drinks or late night food from room service. There had been no suggestion they were inebriated or suffering from lack of sleep, but Brenner knew it could be an issue and wanted it eliminated from his checklist, just in case someone raised it later. After three months, Brenner's subgroup had exhausted its list of chores without uncovering anything significant. They were at a dead end.

But now the Human Factors Group was born again with a stellar membership, mostly Ph.Ds. Having taken five years just to earn a college degree, Robert Sumwalt felt "like a David among all these Goliaths," he would remark afterward. "If the pilots were at fault, these guys were going to find out for sure." Brenner needed every diplomatic skill almost from day one as Graeber, the Boeing-nominated psychologist, emerged as a powerful force on the group. Sumwalt imagined he was watching a contest of intellectual gladiators as Graeber moved quickly to put his stamp on matters.

Graeber launched several proposals for investigation at the group's first meeting, and they were similar to some of the comments in Purvis's controversial letter. He wanted a detailed examination of the cockpit voice recording to determine if the captain had taken control of the airplane, and if he had, if it was done in a proper fashion. There was a discussion of whether one of the pilots should have said "I have the aircraft," as is laid down in the flying manuals to confirm that he was the pilot flying it at the onset of the emergency. Could this possible departure from protocol be evidence of sloppiness on the flight deck? Was there evidence that the pilots allowed themselves to become disoriented and perhaps, with a lethal combination of left pedal and left aileron, tumbled the aircraft out of the sky? To illustrate his points and show where he was coming from, Graeber introduced two similar case histories.

On a foggy, cloudy day in March 1994, Captain Gary Higby was in the cockpit of a Southwest Airlines Boeing 737 flying into Oakland International Airport. His co-pilot was flying the aircraft, letting the autopilot handle the control wheel steering, while Higby went over checklists and communicated with the tower. They had been cleared to land when, at 1,500 feet, his co-pilot let out a bloodcurdling scream. The aircraft had just been slowed down to about 135 knots with flaps set to the 30 position. "What's wrong?" Higby shouted. The co-pilot did not respond but appeared to be staring wide-eyed at something in the fog outside. Higby noticed that the man's back was arched, and his first reaction was to imagine that perhaps he had been electrocuted. Then the man screamed again and clutched at the control column. At the same time, the airplane started to go into a right roll, and Higby felt one of the rudder pedals strike his ankle. He instantly disconnected the autopilot and struggled with the pedals to try to straighten them out, but the aircraft continued to bank. Recalling a trick from his Air Force days, he advanced the throttle for the right engine to overcome the force of the rudder and signaled for one of the flight attendants to come to the cockpit. The attendant saw that the co-pilot's right leg was extended rigidly in front of him and his foot was on the right pedal. She undid the co-pilot's safety belt, and some pressure came off the foot on the pedal. This allowed Higby to regain control of the airplane, radio the tower, and warn them he had a medical emergency. The airplane had lost 600 feet in altitude before Higby recovered control, perilously close to the ground. He managed to make a perfect landing in Oakland, where medics revived the co-pilot, who had been having a seizure.

The second incident occurred fourteen years earlier. Don Widman was the captain of a Frontier Airlines 737 making its final approach to Cheyenne, Wyoming. It was daylight and the weather was clear, and the co-pilot, who was flying, decided to add 10 knots to the airspeed to compensate for possible wind shear. At 800 feet, Widman noticed the speed starting to creep upward, and at about 600 feet it had reached 160 knots. "We're too damned fast," he called out, then announced a go-around, a procedure whereby the landing is aborted and the aircraft makes a circuit until it is ready to try again. But when he turned to his co-pilot, the man appeared dead, with a blue-purple color around his lips and his hands hanging limp. Worse, the aircraft was yawing and rolling to the left, and they were now probably less than 500 feet from the ground that was rushing up to meet them. Widman fought the roll with left aileron and advanced the throttles for the go-around climb, but found it impossible to straighten the aircraft. So he settled for a turning climb and went almost in a full circle before he reached 1,500 feet and regained a bit more control over the aircraft. When a flight attendant rushed in, she spotted one of the co-pilot's legs extended rigidly in front of him, pressing down on the left pedal. She pulled it off the pedal and Captain Widman regained control. His co-pilot, too, had had a seizure.

The stories of Widman and Higby were just two of a number of interesting cases uncovered by Graeber when he trawled through the literature for precedents of in-flight incidents that might be compared to Flight 427. Not only were these good examples of how an aircraft might be accidentally propelled into a dangerous dive, but Graeber was also struck by the admission of both men that they had been startled, and that had delayed their ability to fully understand what was happening to their aircraft. Widman had not realized that the rudder pedal was down until the flight attendant drew his attention to the co-pilot's leg. Higby reckoned that being startled delayed his response by two, perhaps three seconds. Could the crew of Flight 427 have been startled when they flew into the wake of the Delta 727 ahead of them? From his NASA days, Graeber was familiar with earlier research on the effects of being startled, and he dug out the papers.

Farnborough, outside London, is the home of the Royal Air Force's Institute of Aviation Medicine, where in 1972 a program was started to investigate the psychological background to military flying accidents. By the time the program ended in 1988, 148 crashes had been profiled in a bid to discover if closer scrutiny of the personality traits of pilots and the stresses acting on them could help prevent accidents. For example, was a pilot about to get married or having an argument with a superior officer more likely to crash his plane? There were found to be some, albeit minor, correlations between stress and crashes, but even more significant was the rate of what the researchers called cognitive failures due to either under-arousal or over-arousal. In other words, they discovered pilots' actions sometimes failed to match their intentions when the unexpected happened, like getting lost, or a bird strike, or engine failure. Under-aroused pilots, often also fatigued, lacked the ability to react properly. Over-aroused pilots overreacted. Either way, both types made often-fatal mistakes. In ten crashes, for example, cognitive failure was blamed for forgetting to lower the wheels before a landing.

A study by Richard Thackery of the FAA's Civil Aerospace Medical Institute found that in lab experiments, subjects who had been unexpectedly startled by a sudden loud noise took longer to complete simple tasks. Although conducted with volunteers, not trained pilots, this research, along with the Farnborough study, was to be quoted again and again by Boeing for three years following the crash of Flight 427 as evidence that the pilots could have reacted improperly to being startled by pushing the left rudder pedal—either inadvertently or because they were startled into making a wrong decision.

Graeber also raised the question of the crew's medical records and proposed they be checked to see if either man had some condition that might affect his ability to fly which had hitherto remained undeclared. Normally, a detailed check of medical records was not a routine part of a crash investigation because personnel records in the form of sick leave sheets usually told investigators as much as they needed to know. In addition, Graeber made a proposal to minutely examine the pilots' training records to see if their instructors had noticed any predisposition to using the wrong controls. There were also suggestions to follow up reports in which other pilots had successfully survived encounters with wake turbulence and other loss of control incidents to see if they had done anything differently.

Graeber also produced a list of events in which the misuse of the rudder had led to trouble. In 1985, a Midwest Express DC-9 taking off from Milwaukee crashed, killing 31 people. Investigators discovered that 4 or 5 seconds after takeoff, the right engine had failed, but instead of applying left rudder as needed, right rudder was used with the elevators, and the aircraft stalled and crashed. And there was an Air National Guard C-130 that in February 1992 went into a roll and crashed at Evansville, Indiana, killing 16, when the pilot used the wrong rudder pedal. There were other cases where, fortunately, the aircraft recovered safely but where a pilot had applied rudder when no rudder was required. Less happy, however, was the outcome of a Sahara Airlines training flight at New Delhi in March 1994, where the instructor throttled back the left engine to simulate engine failure at takeoff. The trainee pilot, according to Boeing's reading of the flight data recorder and cockpit voice recorder, mistakenly put in a left rudder instead of right rudder. The instructor took over and reversed the rudder, but it was too late and the airplane crashed, killing all aboard and 4 people on the ground.

Another suggestion by Graeber was that the Flight 427 pilots, as their aircraft started its roll through the sky, had somehow become confused through a process known as vestibular disorientation and made a bad situation even worse. A pilot who is flying without visual cues to the world around him can easily become disoriented. Pilots flying manually in fog or on a dark, cloudy night can find themselves upside down if they fail to keep a close eye on their instruments, especially the artificial horizon. A dangerous situation often arises from flying low over a calm, featureless sea in haze that gives the sea the same color as the sky, conditions believed to have led to the death of John F. Kennedy Jr. off Martha's Vineyard in July 1999. An added difficulty is the effect of rapid turns and high G forces on part of the inner ear, the vestibule, and the semicircular canal, the delicate organ by means of which we keep our balance and relate to the spatial environment around us. Disorientation is often worse after a period of rapid acceleration, especially during nighttime takeoffs.

In Britain during World War II, Royal Air Force chiefs were perturbed by a series of crashes of Spitfire fighters that appeared to take off normally on dark, cloudy nights, then flew straight into the ground. Researchers discovered that the acceleration forces of a Spitfire were felt mostly on the back of the pilot's seat, and lacking orientation cues from the dark outside world, the pilot interpreted that as meaning that the aircraft was pointing steeply upward and a stall could be imminent. The often fatal response was to push on the control column to bring the nose down from what the pilot thought was an alarming angle, when, in fact, the aircraft had been climbing normally and was still close to the ground. Lacking visual references to tell him he was now descending and no longer climbing, the pilot would fly his fighter straight into the ground.

Fighter pilots are also particularly vulnerable to a phenomenon whereby the ear gives a false reading after coming out of a tight turn, and they imagine they are flying straight and level when their aircraft may have developed a sinister banking turn. Another disorienting phenomenon occurs when a pilot, glancing rearward out the side cockpit window during a steep banked turn at night, sees the moon underneath his raised wingtip. The moon is exactly where it should be, but the pilot assumes he has inverted, and in attempting to right the aircraft actually turns it upside down. This phenomenon is even more pronounced when an already disoriented pilot has, unaware, entered a banking turn and panics at seeing the moon beneath a wingtip when he imagined he was flying level.

Short of barreling around in a 737 themselves, the best way for the Human Factors Group to check out the vestibular disorientation theory was in a full vertical motion simulator. NASA maintains the world's largest such simulator at its Ames Research Center at Moffett Field, California, and has used it for training astronauts. Not only does it simulates high G forces, it actually reproduces them by rising and falling farther and quicker than any conventional simulator. It rises and falls rapidly through as much as 60 feet and can also move 40 feet fore or aft, plus a more limited movement from side to side. In addition, it swivels and tilts just like a conventional simulator.

The FAA representative on the Human Factors Group, Phyllis Kayten, was an FAA liaison officer with NASA at Moffett Field, and used her contacts to get some time on the simulator. She also suggested that Malcolm Cohen, a NASA expert on pilot disorientation, should sit in on the sessions. Cohen's specialty was the effects of acceleration and G forces on spatial disorientation. He encountered a carbon copy of the Spitfire effect in the late 1960s when he helped investigate a rash of nighttime crashes by A-7 aircraft that flew straight into the sea after taking off from the decks of carriers. The Spitfire effect now had a name, the Somatographic Illusion. Because visual cues were missing, the predominant cue was the pressure from the seat back, and it told the body that it was inclining toward the vertical. Like the Spitfire, the A-7 was vulnerable to a stall, so the pilot's instinctive reaction was to push the nose down, a fatal mistake in many cases.

NASA's vertical motion simulator was programmed to replicate Flight 427's final moments, including the G forces. Cohen sat in with each member of the group. A synthesized version of the cockpit voice recorder tape was played in the background. The NTSB had contacted the pilot of the Delta 727 that preceded Flight 427 into Pittsburgh, and he had said that the day was clear with a very evident horizon, a factor that was reproduced in the simulator's screen. When it was Brenner's turn in the simulator, he realized he could pick out Pittsburgh on the cockpit video display in a fraction of second. His eye told him exactly where he was—exactly were the sky was and where the ground was. He thought the motions were not violent, but were very gradual.

Phyllis Kayten had been apprehensive about the simulator ride. Despite working in the aviation industry most of her life, she was not a pilot and was a nervous flier. Her Ph.D. dissertation had been on how stroke patients cope with processing speech sounds. Her introduction to the NTSB, her first professional encounter with the aviation industry, had been through her father, a veteran of NASA and the aerospace manufacturer Martin Marietta. "I had convinced myself I'd be thrown out of my seat with this huge drop. They simulated it until just before the airplane had been absolutely uncontrollable, yet there was nothing sudden or jarring in that whole ride, we were very surprised with that. On the CVR synthesis you hear the captain saying 'I see the Jetstream, zuh ...' then the pilot says 'Whoa ...' and you can see he said it out of a sense of mild surprise."

"I am fairly confident that pilot disorientation was not a major causal factor in the crash," Cohen wrote to Brenner afterward. His letter continues:

> ... In my opinion, the accident situation did not provoke any obvious evidence of factors that are normally associated with disorientation due to abnormal vestibular stimulation.... This accident happened during clear, daytime, visual flight conditions where there would be ample opportunity for visual information to override any vestibularly-induced disorientation. The motion of the aircraft, from the initial encounter with the turbulence to the point where it probably was out of control and no longer recoverable, did not display obvious evidence of the type of acceleration that would be conducive to disorientation. Rather, except for the initial upset from the turbulence, the motions of the aircraft appeared to have been relatively gradual, supra-threshold, and nearly continuous. Under these circumstances, I believe the pilots would have experienced little difficulty in maintaining an accurate perception of their orientation, even during any brief periods when they may have lost sight of the horizon due to the pitch down attitude of the airplane. In addition, perturbations of the flight path generally appear to have been followed by verbal comments from the pilots, indicting that they were fully aware of their trajectory, and that they were not able to change it. On balance, there does not appear to be any compelling evidence to conclude that the pilots were disorientated, nor is there evidence to believe that they applied incorrect control inputs in an attempt to overcome their disorientation, and thereby caused the accident.
>
> Whether the control inputs were appropriate or inappropriate, it is most unlikely that they were caused by pilot disorientation. Thus, although I cannot completely exclude the remote possibility, it does

> not appear at all likely that pilot disorientation due to abnormal vestibular stimulation provided a major contribution to this accident....

To put more plainly, the pilots knew which way was up.

Both Boeing and the Air Line Pilots Association had a stake in what the Human Factors Group would decide. Boeing, through Graeber and Carriker, would strongly push the pilot error theory, while ALPA continued to maintain there was something wrong with the aircraft and the pilots had behaved impeccably, even though Robert Sumwalt, its representative, tried to declare neutrality at every turn. Brenner walked a diplomatic tightrope, allowing each side to have its say and suggesting further avenues for investigation, but he soon came under fire.

On June 14, 1995, Graeber and Carriker jointly penned a letter of complaint to Brenner, copied to John Purvis, about his management of the group. This followed a June 6 meeting when the group reached a conclusion about the use of the rudder: "There is no way we can conclude for certain that the crew did or did not put in rudder input." Carriker and Graeber wrote:

> ... It seems evident from our recent meetings that too often the discussion by our group lacks focus. Our discussion usually breaks down because one or more participants in the group challenges the work performed by other groups in the overall investigation. For example, one member of the group will say that the airplane 'entered the wake vortex' and another member will respond that it is not known whether the airplane entered a wake vortex. This discussion occurs regardless of the fact that the subject of wake vortex encounters had been assigned to the Performance Group for an in-depth study. Another example relates to the desire of certain participants in our group to revisit, reexamine, and theorize about airplane system failures that could have contributed to the accident....

This "continuous second-guessing" of other groups' work was, they continued, a major reason why the Human Factors Group had become "distracted from gathering human performance data for analysis by the NTSB." Graeber and Carriker wanted the NTSB to "instruct" members of the group to assume that the airplane flight controls were operating normally, or at least to establish that they were operating in some known way: "The leadership of the

Human Performance [sic] Group must establish the starting point for our group's efforts. Only when we can establish conditions for the flight will we be in a position to evaluate human performance. Without such direction, we will be left with the subjective beliefs of individual participants that take us from one tangent to another, and we will not be able to establish the potential crew operational scenarios needed to facilitate our efforts."

Despite the openness of the investigation and the constant sharing of information among its participants, the Graeber and Carriker letter was not circulated among the other members of the Human Factors Group. (It was, nonetheless, subsequently placed in the public docket, the file the NTSB maintains on all crash probes, which is open to public inspection even while an investigation is still under way.) Brenner appears to have tightened up his act after this incident. According to members of the group, he was firmer in the chair and laid down more specific guidelines for the group's operation, although he was still unable to prevent the squabbling that occasionally erupted between Graeber and others.

Graeber wasn't the only one to complain about Brenner. Phyllis Kayten heard disparaging remarks by another member of the group in a bar after a group meeting. She recalls that she "could see where they were coming from. Malcolm is a wacky sort of guy but we all really liked him. That guy should have been a journalist. He has the ability to sit in a bar and get information out of everybody in it, he could go up to guys I'd never sit near and within minutes he's listening to their life stories. I do remember people complaining that there was no strong leadership on the group and that people were being thrown from one proponent to another. He wanted us all to be friends but he wasn't going to set down laws; it wasn't his style, he's very laid back."

But for Kayten, the real problem was not Brenner's chairmanship but Graeber's insistence on getting his way. Graeber and Kayten had known each other at the Ames Research Center, where Kayten had often ribbed Graeber for his inclination to loudly blow his trumpet about the work he did on circadian rhythm, the natural 24-hour body cycle whose disruption causes jet lag.

At the first meeting of the new Human Factors Group, Kayten groaned inwardly when Graeber arrived with a large pile of folders containing research reports he had amassed. Included among them was a study of a series of car accidents caused by drivers stomping on the gas instead of the brake in a new European model with unusual pedals. Graeber held this out as an example of how the pilots of Flight 427 could have erred. If experienced motorists could end up pressing the wrong pedal, why not airline pilots? Kayten recalls, "There were hundreds of pages of research reports which were not exactly to the point because so few of them dealt with experienced pilot behavior. I remember thinking, this is what happens when you have a whole lot of money to spend. I could only imagine what his budget must have been."

Robert Sumwalt launched an initiative to help focus energies by preparing a task-oriented matrix from which to work. Basically a computerized chart, it broke down the various issues that were the subject of speculation and debate into different sections under headings such as: what is known, what the group will be able to discover, what they might be able to discover, and what will never be known.

Kayten found the new direction provided by Sumwalt a relief. For the first time they had a plan that at least directed their energies toward issues that were likely to reward their efforts. But still she was troubled by the transformation in her former NASA buddy since he went to work for Boeing. Had he become what the professional crash investigator hates, the guy possessed by a theory?

Chuck Leonard, Operations Group chief of the Flight 427 investigation, watched the increasingly heated debates from the sidelines. Although the fact that he'd been an airline pilot might have inclined him toward the ALPA viewpoint, he was critical of both sides. "I was very unconvinced and I really felt sorry for the Boeing people who were working on that committee.... It became a matter of disgust with me as I listened to their [Boeing's] position during these meetings. It was only because Malcolm Brenner is such a delightful human being and so tolerant and patient that other people would have said 'you're full of it'."

On the other hand, Leonard accused ALPA of hypocrisy in its view that there was a mechanical fault in the airplane. "They continued to fly those airplanes with passengers in them. They didn't blow the whistle because it could have cost jobs for their membership. If they were so convinced, they could have got those airplanes fixed by stepping up to the plate and refusing to fly them. I was stunned, it was self-serving for them to continue to fly the planes. It discredited ALPA as an organization willing to do anything for safety."

There remained a proposal to take the cockpit voice recorder tape to Seattle to be played in the simulator. This had happened in January, but only with a view of attempting to discover if Boeing's test pilots could identify any of the strange thumps and other noises on the tape. On that occasion, the simulator had been programmed to follow as much of Flight 427's final movements as were known. This time, Graeber and Carriker suggested an attempt should be made to use the tape to help "drive" the simulator. The tape, they said, should also be subjected to multi-track analysis of its sound spectra to discover what noises equated to cockpit controls so they could learn how the pilots were flying the airplane.

ALPA had vehemently opposed the previous use of the tape outside of the NTSB's own sound lab, but Robert Sumwalt, its representative on the Human Factors Group, was in favor of it. Not that he expected Boeing's analysis to throw any further light on the events, but he reckoned it might put an end to what was becoming an endless speculative debate about the suitability of the pilots for the job they were doing.

Sumwalt's decision to back a second Boeing venture with the tape, which was binding, did not earn him kudos at ALPA headquarters. Herb LeGrow, ALPA's accident investigation coordinator, was furious. Pilots had consistently opposed the use of the cockpit voice recorder in crash investigations outside of tightly controlled conditions for a variety of reasons. Paramount was the fear that tapes would leak out and be played on radio or TV (since then this has actually happened) and cause immense distress to relatives unexpectedly hearing their loved ones die. A secondary issue was the danger that any relaxation of their strict stance on tapes might lead to airline management insistence that tapes be routinely checked, and they would become an invisible supervisor in the cockpit.

When LeGrow calmed down, Sumwalt tried to get his point of view across. He relayed the facts of the debate about whether the pilots were following standard operating procedures in the cockpit. He related how he had been shocked by the rapidity at which things happened when he had listened to the tape for the first time almost six months earlier. "We just walked out of that room and we were all saying, 'My God, it happened so fast.' People didn't talk about that theoretical stuff about cockpit resource management and standard operating procedures any more, just how quickly things happened. My feeling was that if we were going to simulate this thing we needed to make the fidelity as good as we can, including the sounds the crew were making, to keep it in perspective."

Sumwalt had tied ALPA's hands on the issue, so LeGrow was forced to back down. But LeGrow was also ready to return to the fray later in the year, when Boeing sought a further release of part of the tape to do some more close analysis.

ter 13 ATLANTIC CITY

Above the clouds high over Delaware Bay on September 25, 1995, a little over a year after the crash of Flight 427, Les Berven saw something truly remarkable through the windshield of the Boeing 737 he was piloting. It was something few other pilots had ever seen before, and he likened it to experiencing "an art form." Two iridescent, sinuous, writhing, rotating tubes of smoke snaked parallel through the sky. They swooped, they curved, they looped, just like a roller coaster ride, and yet each one kept a preordained distance from the other. Elsewhere, in calmer air, the tubes tracked in the manner of a slowly sinking, wraithlike aerial railway.

What Berven was looking at were the normally invisible twin wake vortices, rapidly rotating spirals of air, streaming from the wingtips of a Boeing 727 that was miles ahead of him. They were now made visible by wingtip gadgets emitting smoke, as if the 727 were barnstorming at a fairground air show instead of flying high in the air near Atlantic City, and not for the purposes of entertainment.

Les Berven was supervisory test pilot in the FAA's aircraft certification office in Seattle. A pilot since he was 16, Berven studied aeronautical engineering at California State Polytechnic. Before joining the FAA in 1976, he had been a test pilot for 13 years, 8 of them at Edwards Air Force Base. He was qualified to fly every commercial jetliner Boeing ever produced, with the exception of the 707. Now he sat in the cockpit of a USAir 737-300, part of a mission to recreate Flight 427's possible encounter with the wake vortices of a Delta 727 just seconds before it crashed, and to determine if there was anything deadly about wake vortices that could have led to the downfall of the jetliner.

Also gazing fascinated at the unfolding scene was Mike Carriker, a senior Boeing engineering pilot for the 737 and a member of the reconstituted Human Factors Group investigating the crash of Flight 427. He sat next to Berven in the cockpit. The books were all wrong, Carriker was thinking. He was accustomed to seeing representations of wake vortices as either expanding funnel shapes or else frozen ropes in the sky. They were not like that at all.

Most surprising was their durability as they hung in the sky almost unchanged for four, even five miles behind the 727 that generated them.

The Pittsburgh crash had a dramatic impact on Berven. He had been on the FAA team responsible for flight-testing the prototype 737-300, an identical aircraft to Flight 427. In 1984, when Boeing was seeking official approval for the 300 series design, Berven was convinced he was dealing with an incomparably reliable airplane. It went through the normal routines like a dream. And when he and the FAA test pilots had pushed the envelope, it responded beautifully. They flew it slower than it was designed to go, closer to a stall than any airline pilot would take it. They flew it faster than it was designed to go, and they put in extreme control commands, like full rudder, to make sure that the ailerons could be deployed to recover level flight. They even disconnected a fail-safe mechanism and deployed midair reverse thrust on an engine to make sure the airplane could safely land even if that happened. When Berven examined the flight data recorder printout after Flight 427 crashed, he decided that only one thing could have pushed the airplane into such a maneuver, and that was the rudder. But how?

Other pilots on Berven's team at the FAA were as shocked as he had been and discussed various ways the accident might have happened, but nothing seemed to fit the actual circumstances. One day in September 1994, about a week after the accident, one of Berven's pilots was sitting in the M-Cab flight simulator at Boeing, waiting for the technicians to complete the installation of a new type of pilot display he was to test. While he waited, he tried some maneuvers to see what the effect of the rudder would have been at 190 knots and Flaps 1, the configuration of Flight 427 just prior to the pilots' loss of control. First, he applied the rudder slowly and opposed it with the aileron on the opposite wing. The "airplane" did what it was supposed to do and held steady. Then he flew it at a lower speed, and the rudder had a greater effect. Flown faster, the aileron had a greater effect. He was surprised, however, when he changed the rate of applying the rudder, pushing the pedal much more rapidly than before.

With the autopilot flying the "airplane" at 190 knots, he pushed the right rudder pedal firmly all the way down. It took a little under 3 seconds to reach that position, but what happened caught him off guard. Before he had time to hit the switch on the control wheel that disconnects the autopilot, the simulator rolled to the right through 50°, and the nose was angling toward the ground. He pulled back on the control column to get the nose up and turned the wheel to the left to stop the roll, but to no avail. The "airplane" continued to roll until it was somewhere between 120° and 140°, almost upside down, before he canceled the maneuver. He immediately reported what had happened to Berven.

Berven was alarmed. Rapid rudder movements had not been a part of normal flight testing when the airplane was being certified. Hardovers were tested by slowly applying pressure to the rudder pedals—doing it fast had simply never occurred to anyone. A team was assembled and a four-hour test session in the simulator followed, with the same disturbing result. Especially worrying was that the maneuver allowed almost no reaction time. Most tests conducted by the FAA before approving a new airplane design assume that it will take a pilot up to 3 seconds to recognize what has happened before taking corrective action, and in that time the airplane should have rolled no more than 60°. The scenario now unfolding allowed very little time before the airplane was upside down.

Later in September, Berven became involved in helping Tom Jacky's Aircraft Performance Group attempt to recreate Flight 427's final moments in the same simulator. Again, he found that the faster the rate of rudder movement, the more rapidly the "airplane" rolled. For example, a rudder moving at 0.5° per second caused the airplane simulator to roll slowly through 70° in 17 seconds. But at 2.5° per second, the airplane rolled the same amount in less than half the time. At the maximum stomping-the-pedal-to-the-floor rate of 5° per second, the airplane simulator rolled through 70° in little more than 2 seconds. The 2.5°-per-second rudder movement in the simulator was, Berven reckoned, the closest he could get to reproducing the airplane maneuver registered by Flight 427's flight data recorder at the start of its departure from normal flight. Flight 427 had rolled violently to the left, and its nose also started to turn left. Then, the aircraft had rolled almost onto its side. As the rolling motion increased, its nose started to point downward until it was upside down and twisting toward the ground. In the next 10 seconds, it completed a full 360° roll, still pointing downward. To be certain, Berven suggested that the FAA ask Boeing to run a series of instrumented flights to calibrate the simulator settings. He especially wanted a calibration during what is known as "steady heading sideslip," when the rudder is being used to "skid" the airplane sideways, and the opposing aileron is used to keep the heading steady and the wings level. It would be several weeks or more before Boeing engineers extracted all the data from the calibration flights and applied it to the simulator, but Berven thought the airplane performed as expected. There was just one exception. At 190 knots, the rudder on the airplane was going 2° farther, to about 20° deflection, compared to the 18° calibrated on the simulator.

Berven was also involved in flying the wake turbulence encounters in the Boeing simulators, which he thought closely approximated wake encounters he had experienced while actually in the air. No two encounters are the same, and he guessed that they varied according to how an airplane hit the wake. A crossing at right angles in a real airplane felt like a strong, sharp jolt, and it was accompanied by a sound like someone was walloping the bottom of the airplane with a baseball bat. In the simulator, entering a wake from the side at a shallow angle was even more dramatic, with the simulator rolling up to 30° but stabilizing after a couple of seconds as the airplane was ejected—spat out, he would say—from the vortex.

All of Berven's previous wake encounters had been unexpected. Now, for the first time in his career, a year after the crash of Flight 427, he was deliberately seeking one. The normally invisible wake was plain to see. Two slim pods that looked like rockets on a fighter-bomber were clipped beneath wingtips of the 727, which was owned by the FAA. The pods contained miniature heaters that burned Clovis oil, emitting a dense gray-white smoke. As it spewed out, the smoke was sucked into the vortices and marked them for miles behind.

Berven's group had finally secured a test airplane thanks to USAir, which agreed to lease them a 737 for the week or so they needed it. Were it not for that, the tests would have been abandoned, as all efforts to secure an airplane on the open leasing market had failed. (No companies, it seemed, were prepared to rent out an airplane for as little as a week, and agreeing to the longer-term leases would have proved prohibitively expensive.)

Over the next eight days, different pilots would fly the 737 into the 727's wake from a range of angles to see what the result would be. Their airplane was highly instrumented and was also equipped with seven video cameras, including one on the tail. The cameras enabled investigators in the Aircraft Performance Group to precisely calibrate the movement of the airplane with its position in the wake vortices. The airplane was also fitted with special quick-access flight data and cockpit voice recorders. It was not alone in the sky: a T-33 jet trainer supplied by Boeing accompanied the flights. One of the test's objectives was to mirror as closely as possible the weather conditions that had prevailed on September 8 a year earlier. For two days, a NASA OV-10 research airplane monitored atmospheric conditions and made its own penetrations of the vortices.

For safety reasons the initial test, on September 25, 1995, was performed at a fairly high altitude, above 18,000 feet, just in case the encounter precipitated something unexpected and the airplane needed the altitude to recover. They headed out over the Atlantic, where the weather was calmer, and practiced lining up with the smoke trails. Then they gingerly stuck one wing in, then another, and the response was nothing to worry about. Flying beneath the vortices, they stuck the tail fin in, and the resulting shimmy was easy to recover from. Emboldened, they started to fly through the vortices from above, from below, and from each side. "Hey, this is no big deal," thought Berven. The sideways approaches produced the greatest effect on the airplane's flight, but there was no difficulty regaining control, even when the maneuver was performed with the autopilot switched on. After a half hour, the weather started to deteriorate, and it was time for the small aerial armada to return to Atlantic City.

Poor visibility grounded the fleet the following day, but they received the go-ahead for more tests on day 3. The flights continued for the next five days at gradually lower altitudes until they were at the 6,000 feet, where Flight 427 was flying when it may have met the vortices of the Delta 727.

The experience changed Berven's attitude about the phenomenon of wake turbulence. Previously, he thought that a wake encounter of the magnitude he was experiencing in the Atlantic City flights had been a warning that there was something even bigger around. As he explained to Tom Jacky at a second public hearing, held in Springfield, Virginia, in November 1995, "Typically, you'd be flying along and all of a sudden, whoosh, you'd be rolled up a little bit and back out again and you'll say, ah, a wake encounter. And sometimes you'll say, boy, we really dodged the bullet on that one because we must have been very, very, close to this other big vortex over here because that wasn't very much. So you say, wow, got out of that one again."

Now, at the second public forum, he was willing to state that there was no big vortex, at least when following an airplane the size of a 727. One would usually get 10° or 15° of rapid roll. "That's as bad as it gets, you're never going to see anything worse than that." To get a heavier roll, up to 30°, the pilots had to almost force the airplane where it did not want to go, into the vortex at a shallow angle from slightly below.

Four miles back, the vortices were always found 300 feet below the 727, an almost precise correlation with the Flight 427 re-creation. Berven was also struck by how these perfectly formed things 4 or 5 miles behind the 727 were still tracking together, still rotating, even when they were being tossed about by atmospheric turbulence. Flying up and down between the two vortices produced very little roll, just a few bumps. But even when the airplane rolled steeply, none of the encounters produced much by way of pitch or yaw. And yaw, going by the experience of Flight 427, was what the investigators had been expecting. The flight data recorder showed a yaw, just at the moment when the airplane would have crossed the wake of the Delta 727. But in the Atlantic City tests, there was no yaw.

Nothing about the maneuvers was startling, Berven concluded. The autopilot did such a good job that the airplane didn't roll more than 10°. The steepest rolls were those when the crew deliberately took no action to counteract them. Because the vortex actually ejected the airplane, rolls lasted only a couple of seconds at most. Berven was also pleasantly surprised that the way the Boeing simulator was programmed for vortex encounters was surprisingly accurate.

"The roll angles we got out of the simulator before we did this test were almost exactly the same as we got in the airplane, the only difference was that in the real world, they were a lot closer; yet it had the same effect, it just happened quicker," he explained to Jacky at the second hearing. Berven had taken advanced aerobatic training, and Jacky was curious to know if that had aided his ability to deal with the vortices. "No," Berven replied, "the angles and the rates were so small and so slight that it was nothing approaching any kind of aerobatics maneuver or anything you would consider as an unusual upset or unusual attitude."

At one stage, Berven positioned the airplane so that a vortex went straight through one of the 737's engines. According to pilot lore, a strong vortex is capable of producing a "flame-out"—the shutting-down of an engine—but apart from a strange noise like someone blowing over the top of an empty bottle, nothing happened. In fact, as they watched the instruments, the engine continued to produce power at a constant rate.

The Boeing pilot, Mike Carriker, found nothing frightening about the encounters, but then, he reasoned, he had been mentally preparing for them. But he was surprised by the unpredictability of the airplane's responses, often because of the way the wake might bend. As he explained to Jacky at the second public hearing, "You think you've got it all set up, you think you've got it and this is what's going to happen and it doesn't. You think you're not going to have a roll and you get a bigger roll. You think you're going to have a bigger roll and you get a smaller roll." He was also impressed with the ability of the autopilot to dampen out the rolling motions as the airplane passed through the vortices. And he was in agreement with Berven about the lack of yawing motions: the wake vortex encounters had almost no perceptible effect on the heading of the airplane, only on its rolling motion. On one flight, where he was attempting to keep the fuselage within the vortex core, a difficult task, Carriker experienced a 60° roll, although the average was less than 30°. On that occasion, the autopilot had been disconnected and he had deliberately taken his hands off the control wheel to see what was the worst the vortex could do.

Was he surprised or startled by the encounters? Jacky wanted to know. Carriker's answer was a definite "No." Did he ever feel he might lose control of the airplane? "No." Did he think other crews might overreact to an encounter with a wake vortex? "I don't know, I think that's tasking my Human Performance Group knowledge and I don't really know how to answer that question."

But Boeing's John Purvis wanted to tease that line of questioning out further, to his employer's advantage. When he asked Carriker if wake encounters had the potential to be confusing, and Carriker replied, "Not to me," Purvis immediately followed with, "To other pilots, possibly?" Carriker replied, "I think so, because in pilot reports one-third of all uncontrolled airplane actions are related to wake turbulence events." Carriker, the cool professional pilot, had just slipped into Human Factors mode again, Boeing style.

The wake vortex tests might not have proved much in the way of pilot reaction, but they did show that, on their own, wake vortices were an unlikely direct cause of the Flight 427 crash. The data showed that the wakes on their own could not have rolled over a perfectly functioning 737.

Berven was also impressed by the sounds the vortices made when they hit the windshield. It sounded as if the airplane were struck by a stick. The strange thing was, you could hear the noise but you couldn't feel it. There was no jolt. Jim Cash, the NTSB's acoustic expert, was also interested in the noises heard by the pilots. Cash was an electrical engineer and a former US Air Force fighter pilot who once flew F-4s. He was deeply frustrated by his inability to decipher some of the unusual clicks and thumps recorded on the Flight 427 cockpit tape. After 32 minutes of normal flight, there were what appeared to be three faint thumps, and then the plane was out of control and heading for the ground. Dozens of pilots and engineers had also listened to the noises and declared themselves equally baffled. Now Cash, who had taken over from Albert Reitan in attempting to decipher the sounds on the tape, had rigged the USAir test plane with a quick-access cockpit voice recorder. If the three thumps were sounds normally made in flight, or even in a wake vortex encounter, he would soon know. On the other hand, if the Atlantic City tests were unable to duplicate the noises, he was still in a quandary.

As a member of the Cockpit Voice Recorder Group (in addition to his role on the Human Factors Group), Mike Carriker had often listened to the cockpit tapes from Flight 427. When Cash asked if the noises heard passing through wake vortices were the same thumps he had heard on the tapes, Carriker said no. The thumps didn't sound anything like the "whooshing" noises he had heard in the vortices. Nevertheless, Cash unloaded the quick-access cockpit voice recorder tape from the test plane, and in his hotel room that night played it on a portable tape deck. To his amazement, he heard the same sound that he had so often heard on the Flight 427 tape, just before the airplane commenced its fatal dive. But when he played it for Carriker, the latter said no, that was definitely not what he heard in the cockpit. The mystery deepened.

Cash had quite a résumé in using acoustics to help solve mysterious crashes. He was often asked by his NTSB colleagues for help in deciphering what the engines were doing just before a crash, especially on older Boeings with primitive flight data recorders. Though many older recorders did not record engine speeds, over a ten-year period Cash had learned how to analyze engine sound signatures on cockpit tapes. He could tell if an engine was going fast or slow, or if it was powering up or slowing down, and figure out the rate that it was throttling up or down. He could even tell if one engine was performing differently from another or had failed. He became so adept that he could analyze a tape of an airplane rolling down a concrete runway and tell at what speed it was traveling by comparing it to tapes of similar airplanes on the same runway.

Explosions were another important item in Cash's repertoire of acoustic memories, and he had carefully listened to the Flight 427 tapes for the telltale sound of a bomb. All the while that the FBI was combing through the wreckage at the crash scene and in the A1 hanger in Pittsburgh, he was confident that a bomb had not downed this airplane. Later he would learn to accurately distinguish the difference between the sounds of a bomb, a rocket, and a fuel tank exploding.

When the FBI and the CIA decided to plant warheads from US Navy rockets aboard several scrapped Boeing 727s in a giant airplane boneyard in the Arizona desert, they did so because they wanted to see what kind of damage would result, and to compare it to the wreckage of TWA Flight 800. At the time, the FBI was pursuing two theories about this high-profile July 1996 crash off Long Island: either a bomb had been planted aboard, or a stray rocket from a naval exercise hit the airplane. Jim Cash tagged along with his tape recorder because he wanted to see what kind of noise those warheads made.

When he learned that a scrapped Boeing 747 was to be blown up in England as part of a joint US-UK test of a bombproof luggage container, Cash asked if he could lug his recording equipment along. For the container test, he teamed up with an acoustic expert from the University of Southampton doing research on behalf of the UK's Air Accidents Investigation Branch. They were allowed to explode small charges in different parts of the airplane, such as a lavatory, and record the sound. They set off more than 60 explosions in the airplane, some using high explosives, others using aviation fuel. When the luggage container was tested in a larger explosion, they also recorded that. The rear half of the airplane was practically demolished, but the center wing tanks were still intact.

Because these tanks were believed to be the site of the explosion aboard TWA 800, Cash wanted to experiment. They first exploded small charges of plastic explosives to measure the sound signatures. Then they repaired the holes, filled the tank with a mixture of propane and air to simulate a fuel-air mixture, and ignited it. "It was sort of still standing when I left but there was a lot of it spread around the place," he remarked. Those experiments confirmed for him the difference between a fuel explosion and a bomb, which to the untrained ear sound identical. Cash listens for what he calls the rise time, the amount of time it takes for the sound of the explosion to reach its maximum amplitude. A high-explosive bomb might take 0.002 seconds in rise time, while a fuel tank exploding takes at least five times as long.

Cash also measured the time it took for the sounds of explosions to reach the cockpit voice recorder microphones. Sound travels in two ways aboard an airplane, through the skin of the metal fuselage and through the air. A sound traveling through the skin moves at least nine times faster than an airborne sound. If both are picked up on a cockpit voice recording, then the difference between them can be timed and the location of the event causing the sound accurately located.

But these advanced acoustical calculations were in the future. In September 1995, Cash set about locating the site of the thumps he had heard on the cockpit voice recorder tape from Flight 427. Using a similar airplane parked on the ground, he started the cockpit voice recorder, then walked around the airplane banging the fuselage with a rubber mallet, carefully noting the time and location of each strike. Then he spent days in the NTSB lab playing the tape at a tiny fraction of its normal speed, measuring the difference between the time of the airborne sound and the faster skin-transmitted sound. The two could be easily distinguished because skin-transmitted sound has a characteristically lower frequency than air-transmitted sound. Cash scrutinized the timing of the mysterious thumps on the original Flight 427 tape, using waveform printouts to accurately place the onset of each sound. Eventually, he found a match: striking the fuselage of the parked airliner just 16 feet back from the cockpit area microphone, around row 1 or 2 in first class, produced the same time gap, measured in thousandths of a second, as that between the air- and skin-transmitted sounds aboard the crash airplane. He now knew roughly where the sound had come from. The question remained: what had caused it?

Now, after the Atlantic City tests, Cash had seven videotapes of the wake vortex encounters shot from different parts of the test 737 to work with, plus the footage shot from the T-33 trainer that had acted as an observation airplane. He calibrated the videotapes with cockpit sound recordings until he had an exact match. Then, frame by frame, he moved the video and sound tapes in tandem until he reached the spot where the thumps started. As the airplane turned into the vortex smoke trails, he witnessed something remarkable. The smoke trail split as the fuselage entered it. On the side of the airplane, facing the inside of the turn, there was a steady stream of smoke along the fuselage. On the other side, facing the outside of the turn, the smoke did not stream steadily but moved a few feet outward, then curled back in to hug the fuselage for the remainder of its length. The point where it rejoined the fuselage was the exact spot where Cash had determined the thumps originated. As Cash himself described it, "It appeared to be a flow separation, then the flow reattaches back to the fuselage. It was almost like a hammer blow, and it happened where the pointy end becomes a tube, around the number one cabin door, where it is no longer tapered and becomes a straight fuselage."

Cash's findings also explained why Carriker said the sound he heard in the cockpit differed from the sound on the tape. The whooshing noise he heard in the cockpit was a higher frequency airborne sound. But the cockpit area microphone, fastened to the airplane's structure on the overhead instrument panel, was picking up a combination of higher frequency sound transmitted by air and lower frequency sound transmitted by the structure. Had Cash been able to strip out the sound transmitted by the structure, he had no doubt that Carriker would have recognized the remaining sound as what he heard in the cockpit.

But there was still another mystery. Although Flight 427's engines, as recorded on the flight data recorder, appeared to be running normally and were not being throttled up or down as the accident commenced, the cockpit voice recording told a different story: the sound of the engines increased in volume. Run through a computerized waveform analysis, the pitch of the sound remained unchanged, but its loudness increased. So why did the sound get louder? One possible explanation was a change in the way sound was being transmitted.

Cash had heard changes in engine noise like that in 1989, when a cargo door flew off a United Boeing 747 as it climbed out of Honolulu, Hawaii, tearing a gash in the fuselage and causing nine passengers to be blown out of the plane to their deaths. He also heard the same changes in 1988, when a section of roof peeled off an Aloha 737, also over Hawaii. Did a door fly open aboard Flight 427 in midair, allowing more noise into the airplane? Did something fall off the fuselage, creating a hole that admitted more noise into the interior? One of the consequences of a door opening or a section of fuselage disappearing is an increase in wind noise, but there was no indication of this on the

Flight 427 tape. Did a bird strike create a hole in the plane? Whatever it was, if it led to the engine sound increasing, could it also have caused the plane to go out of control?

What, if anything, had occurred was a mystery and might have remained one forever except for the Atlantic City wake vortex tests. Because there it happened again. One day after the wake vortex tests had been completed and the FAA's 727 was returning to base, Les Berven ran a separate series of maneuvers with the 737. He was trying to fill in some of the gaps in his knowledge of the way the Boeing 737 behaved in the air. If the rudder is jammed at a certain position, the pilot can overcome this by using the ailerons on the wings. But nobody knew exactly how the 737 behaved when the rudder was applied and then countered by what pilots call opposite aileron. If the jammed rudder is causing the nose of the airplane to swing to the right, for example, using the left aileron can overcome it and push the airplane back on course. However, even though it may be going from A to B in a straight line, the airplane would not be lined up with its course but would crab sideways, or "skid," through the air in what is called a "steady heading sideslip."

Cash later dutifully listened to the tapes of these maneuvers, not expecting anything dramatic, but was surprised to hear the engine sounds increasing in volume at certain moments when Berven was putting the airplane through its sideslip paces. Then the noise would subside back to normal levels. He had been on board during those tests and could not recall anybody playing with the throttles. When he got back to Washington, Cash eagerly obtained a copy of the flight data recorder output from that day's flying. He correlated the two records until he had an exact match in time, then played the tape through slowly and noted, according to the flight data recorder, what the test pilot had been doing at the precise moment the engine noise increased in volume. He excitedly realized that Berven had been applying the rudder to yaw the airplane prior to countering the yaw with ailerons to produce the desired sideslip. When the sideslip became steady, the engine noise subsided to normal values.

Cash encountered the increase in sound volume on more than one occasion, but not during every rudder movement. During more gentle rudder movements, there was little or no change in volume. It was most noticeable, and closest to that heard on the Flight 427 tape, during a more vigorous movement that swung the rudder out between 7° and 14°. In the end, though, Cash found the increases in engine noise were most alike when the *rate* of rudder increase was about the same. The extent to which it moved had less to do with it.

Cash developed some theories about why this should happen. "Look at the airplane and you can see that a large portion of the sound is going straight out of the engine because when they design it, they try to minimize side noise because that's how they can get nailed on takeoff noise measurements. So they shoot most of the noise out the front of the engine because they are trying to suppress side noise, which also affects passenger comfort. That's fine as long as the airplane is flying normally. Now if you yaw it, that sound impinges differently onto the fuselage. We appeared to be hearing some of the sounds of the first stage fan momentarily come back onto the fuselage."

So nothing had fallen off the airplane or come unlatched in flight. The increases in engine noise came just as the accident sequence was underway. The first increase occurred 170 milliseconds (0.17 seconds) after the second set of thumping noises began. The next occurred 1 second later, just before the captain yelled "Hang on!" for the first time. Investigators had been unsure of what the rudder was doing because of the paucity of flight data recorder information. But Jim Cash had come up with a way of finding out.

hapter 14 JIM HALL STEPS IN

Unlike his predecessors at NTSB, Chairman Jim Hall took a very strong public stance on the issue of flight data recorders, typical of the approach he would take with other safety issues like child seats in automobiles and airplanes. When he met relatives of people who died in the Pittsburgh and Roselawn crashes and learned of the shabby treatment some of them received from the airlines and public authorities, he made himself and the NTSB their champion.

Hall became convinced that the relatives of plane crash victims needed greater protection vis-à-vis the airlines after the distressing stories he heard. Relatives phoning the 800 number given out by an airline in the wake of a crash were often unable to get through, leading some to wonder if the lines were manned at all. Eight hours after the crash of Flight 427, many relatives had still not been officially informed by USAir that a loved one had died, even though he or she was booked on the flight and had not phoned or returned home. The night of the crash, relatives who traveled to the Pittsburgh airport seeking information were kept in an isolated room to which others, even clergymen and mental health counselors, were denied access.

The sparse and sporadic nature of the information given by an airline convinced many relatives that the 800 number and the airline's representatives looking after them in the hours immediately following a crash were devices enabling the airline to compile a more accurate passenger list. Airlines appeared more worried about the legal consequences of informing the wrong person, like a relative whose loved one had actually missed the flight and survived, than they were about keeping the genuinely bereaved abreast of the situation. It was as though they believed the bereaved were going to sue them anyhow, and they wanted to make sure they wouldn't attract additional suits they might otherwise avoid.

The length of time it took to remove remains from the site of an air disaster caused further grief to relatives, who were not informed of the forensic processes involved or the mutilated state of the bodies. After Flight 427 crashed, it was two weeks before Robert Connolly's remains were released to his family and a funeral could take place, his twin brother Dennis would recall.

Aside from a jawbone and parts of a torso, there was little else of Robert in the coffin, Dennis later discovered. Remains were often misidentified, and unidentifiable body parts were sometimes left at accident sites. After the Pittsburgh and Roselawn crashes, both USAir and American Eagle buried unidentified remains in secret mass graves, to the distress of relatives who learned that only a portion of the remains of their loved ones were in the caskets they had buried themselves. There was also anger at coroners' staffs, who allowed control of the remains to rest in the hands of the airlines.

Personal effects were mislaid or, in one case, incinerated. After Flight 427 crashed, relatives were not told that personal effects had been discovered in the wreckage taken to the Pittsburgh hanger by crash investigators. Some airlines refused to return damaged personal effects, no matter how important, in the mistaken belief that doing so might cause further distress. "It may seem insignificant, but to us it was very important," said Dennis Connolly. "This was something that was with a loved one at the very end, it was our last contact with them when they were alive."

Airlines tried to control the situation by convincing relatives that they, the airlines, were the appropriate sources of information and assistance after a crash, and there was no point in turning elsewhere for help. In fact, as some relatives discovered, this part of the process was being stage-managed by lawyers. Some grieving relatives discovered to their horror that what they said—believing it was in confidence—to an airline grief counselor was being relayed to the airline's legal team for use against them in subsequent litigation. Jim Hall was told by one woman that she mentioned to a counselor supplied by an airline that she had gone to a therapist during an earlier divorce. This admission was thrown at her in front of a jury to make her look bad, presumably to somehow reduce her financial award.

Dennis Connolly was at first impressed by USAir's generosity when it paid for the funeral of his brother and flew relatives into Pittsburgh to attend. But his admiration soured when he discovered that some of the bereaved had funeral and travel expenses incurred by the airline deducted from subsequent legal settlements. After the crash of Flight 585, United Airlines deducted funeral expenses from the awards made to the crew's families, and the captain's daughter was told her father's funeral could not proceed until she signed legal documents presented to her by airline officials at the funeral home.

When Connolly asked USAir for a copy of the seating plan to find out who was sitting next to his brother when he died, not an unusual request by relatives in the wake of air disasters, he was turned down. Later he discovered by accident that the seating plan was included in the NTSB's public docket. "The only reason I can think for their refusal was because USAir did not want us talking to other relatives. It was to protect the airline, not to help us. They did things that were in poor taste, that were very wrong. We were treated like outcasts, we were not involved in any of the really important decisions." When USAir erected a memorial in the Sewickley, Pennsylvania, cemetery, in which many of the dead were buried, it again failed to properly consult relatives. The memorial simply listed the names of the 132 dead, omitting the fact that they had all died together in the crash of Flight 427.

Connolly joined the relatives' group formed after the crash of Flight 427. Such groups spring up in the wake of most air accidents, initially to provide support for the bereaved, but then frequently branching out into campaigning for improved air safety. The Flight 427 Air Disaster Support League decided to concentrate on two major aims. The first was the purchase of the crash site and the erection of a proper memorial for the dead. The second was a campaign for improvement of the treatment of relatives in the wake of air disasters. The group was one of the more effective ones in part because it had a large concentration of its membership in a single city. (The majority of people in the Flight 427 group were from the Pittsburgh area, so meetings were held more often and were well attended.) To a remarkable degree, it showed real cohesion and unity of purpose in its dealings with the agencies and corporations most deeply involved with the disaster. Before long, the group developed a clear sense of who was on their side.

When the Flight 427 Air Disaster Support League lobbied the Department of Transportation and the FAA, they were politely but unenthusiastically received. Sometimes they felt no more than tolerated. But when they went to Jim Hall, they discovered they were pushing on an open door. He championed their efforts to get new legislation passed that placed the interests of relatives and survivors above those of the airlines. He met with them several times, advised them on a lobbying plan, and set up meetings with key congressmen on Capitol Hill. He even went so far as to offer the NTSB as the means for putting the new legislation into effect. Thanks to Hall's backing, the passage of the Aviation Disaster Family Assistance Act in October 1996 was swift and relatively trouble-free. The new legislation placed the NTSB in charge of coordinating support and counseling for relatives and survivors following an airplane crash. The NTSB, through a newly established Office of Family Affairs, would also provide private briefings for family members on the progress of an investigation, thus ending the airline's monopoly on the flow of information.

But while Hall's rating with the families and the public was soaring, not everybody was happy. Hall had joined the board as an unknown minor Tennessee politician and ended up a high-profile national figure showered with awards by aviation trade and consumer groups. All agreed that he did a superb job promoting the NTSB. But there were also those who believed he was using the NTSB to grandstand. Many staffers were worried that this new concern for the public would only distract the NTSB from its more mundane but vital task of finding the causes of air crashes and preventing their recurrence. Other staffers were worried that they might be forced to shed their professional veneer—and more than just a veneer—of emotional detachment from the human tragedy of a crash by being plunged into closer contact with grieving relatives. Intense feeling were intense feelings, these professionals knew, but they did not always lead one down the path to the truth.

The doubts and grumblings did not, however, deter Hall. He wanted more changes, so more changes were on the way. For one thing, he wanted to deal with the kind of frustration he felt when the NTSB was accused of being timid, overly cautious, on the side of industry, or secretive. In particular, he'd been particularly galled by a September 1995 article in *Newsweek* that quoted two lawyers who alleged that the NTSB knew what had caused the crash but would not say:

> ... The flamboyant Philadelphia lawyer Arthur Wolk calls the wake-vortex theory, for example, "the biggest fairy tale I've ever heard." And he insis ts that the Feds know more than they're saying. "How [NTSB chairman] Jim Hall can stand there and say 'We're still baffled' is beyond me. Ever ybody on the inside of the investigation knows—not believes, knows—i t's the rudder."
>
> For attorney Richard Schaden, pinpointing the exact cause of the rudder malfunction has become an obsession. Last February, he bought the tail section of a Boeing 737 and had it trucked from Tucson, Ariz., to his private hangar outside Denver. He'd already bought a 737 rudder-control unit from Peru. A pilot and aeronautical engineer, Schaden has scrutinized every training manual and technical report on 737 rudders he can get his hands on, and even obtained Boeing's computer data, which he's used to program his own flight simulator. He believes that he knows where the rudder problem lies. "This is the troublesome part, the servo valve," he says, pulling a cylinder the size of a juice can out of his rudder-control unit....

Hall was not the only one stung by such accusations. Around the same time, Bernie Loeb, the director of the Office of Aviation Safety at the NTSB and Tom Hauteur's boss, took a phone call from Byron Acohido, the *Seattle Times*'s aerospace correspondent who would later win a Pulitzer Prize for his coverage of the Flight 427 investigation. Acahido told Loeb that the NTSB knew the problem was the rudder and wanted to know what it was going to do about it.

"Yes," Loeb would recall, "we knew that the rudder had the capacity, if it went full hardover, it had more than ample capacity to cause the crash. But we didn't have any plausible explanation as to how the rudder might have malfunctioned to do it. What you needed was a scenario in which you have a mechanical malfunction which you can explain and which would provide the rudder positions at the right time during the event and the wheel positions at the right time to fly you through the recorded data. In 1995 we weren't even remotely close to anything like that."

Loeb was also coming under pressure from people who had lost relatives in Flight 427. "Some of the families were believing some of the stories being told by some of the attorneys like Arthur Wolk, who said we knew what had happened but that we were in Boeing's pocket and wouldn't own up. On that count, Wolk was just plain wrong."

Privately, Hall was starting to wonder if his staff, despite their best efforts, possessed the ability and the expertise to solve this crash. Normally, the agenda of a public hearing is determined by senior investigators working with the parties, and engineering people like Bernie Loeb and Tom Hauteur would have a major say in setting the agenda. But just prior to the second hearing in November 1995, Hall sprang a surprise. He wanted the issue of hydraulic contamination addressed and nominated a man named Paul Knerr as a witness.

Knerr was vice-president in charge of engineering for a small hydraulic component business in Valencia, California. The company, called Canyon Engineering Products, made hydraulic valves under subcontracts for larger firms. Few NTSB staffers had ever heard of Knerr. His firm produced no components for the 737, and the usual prerequisite for third-party manufacturers to become involved was that their company provided some component for the aircraft under investigation. Knerr had been one of hundreds of people who had offered Hall theories about Flight 427. Hall usually quickly reviewed and passed on these messages to Loeb or Hauteur without paying too much attention, and they often put them in the crackpot file. But Knerr's approach had struck a chord with Hall.

In his search for a smoking gun, Hall was aware that Greg Phillips's Systems Group had microscopically examined the components of Flight 427's rudder power control unit. They looked for scratches and other marks that might be evidence of jamming caused by metal particles or other contaminants finding their way into the 737's hydraulic system. Nothing of the sort was found. The surfaces of the valves and the cylinders that they moved within were polished and flawless, as good as the day they were first machined. But Knerr had familiarized himself with the work of the Systems Group, and he had an interesting suggestion. What if the hydraulic fluid in the rudder system were so contaminated with silt that, instead of acting as a lubricant, it produced enough friction to slow, or even stop, the valve's movements, causing a jam? Knerr's unconventional ideas—and indeed his mere presence—would lead to controversy and division within the NTSB. Yet his involvement also led to unlocking some of the secrets of Flight 427.

Opening the second public hearing—in November 1995, in Springfield, Virginia, 14 months after the crash of Flight 427—Hall directly challenged the critics of his agency:

> ... If indeed somebody has found a "golden nugget" or answer for either one of these accidents, it is odd that he would choose to meet with *Newsweek* and not the Safety Board. Since the accident, I have met on several occasions with representatives of family members, many of whom are in this audience this morning, who lost loved ones on Flight 427. There is nothing I want to accomplish more in my time of service on this Board than to find the cause of this crash. I can only say that if we knew what caused this accident, we would not be expending thousands of hours a month on this investigation. We wouldn't have spent a million dollars last month on a flight test [in Atlantic City]....

Hall added that the NTSB was "looking at rudder issues very hard.... But we need proof to find and cure real problems. We are always ready to consider hard evidence that will withstand the scrutiny of trained investigators, not wild accusations that are eagerly bandied by people looking for a sound bite on television." Hall was distressed, it was clear to those who knew him, and irritated and impatient enough to take the initiative, to push the investigation in whatever direction he thought had a good chance of yielding results.

Though few people at the NTSB had ever heard of Canyon Engineering Products, let alone Paul Knerr, the company and its engineering VP weren't entirely coming out of left field. As a young man, Knerr had worked as a trainee engineer on NASA's Apollo program. During the year just before his appearance at the public hearing, he had worked on and off with Boeing and the FAA in an attempt to come up with a method to continuously monitor the level of contaminants in airplane hydraulic fluid. He sat on a nationwide industry standards body attempting to define limits for hydraulic fluid contamination, the A-6 committee of the Society of Automotive Engineers. (The society's initials, SAE, often preface the classification number of hydraulic fluids destined for applications as diverse as the car in your driveway and a Mars lander.) The A-6 committee was examining contamination and filtration in hydraulic fluids that might contain particles of foreign matter, ranging in diameter from 2 to 100 microns. It had recently been asked by the FAA to follow up some of the hydraulic contamination issues raised by the Michael Zielinski's Critical Design Review team and make recommendations.

The Critical Design Review of the 737's rudder and other flight controls, published in May 1995 by the FAA, listed hydraulic contamination as an area meriting further study. The review uncovered an interesting fact: unlike other plane makers, Boeing did not set any limits on the amount of fluid contamination it tolerated in its airplanes, stating that this was a matter for individual airlines.

Greg Phillips's Systems Group had already tackled the area of possible hydraulic fluid contamination in connection with Flight 427. Samples of fluid found in hydraulic cavities in the crash wreckage were compared to samples taken from other 737 aircraft in service throughout the US. They were analyzed by Monsanto, the fluid's manufacturer, at Tinker Air Force Base labs near Oklahoma City, and at Parker Hannifin. All of the samples were contaminated to some degree, but not to an extent that could cause jamming. The fluid from the crash plane was found to be close to Class 11, a high level of contamination, but not unusual. Each contamination grade represents a doubling of the particulate count over its predecessor. New aircraft were delivered with no greater than Class 9 contaminated fluid. Fluid contamination is not always a measure of carelessness, and it can build up rapidly in even the most carefully assembled and maintained hydraulic system. Moving parts rubbing against each other can shed microscopic slivers of metal, rubber seals can degrade into sludge, dust may be in pipes and reservoirs, and water can condense from any air that finds its way into the system.

In January 1995, Boeing conducted a series of tests to see what might happen to a rudder power control unit loaded with hydraulic fluid contaminated to an even greater degree as Flight 427's. Over an eight-day period of continuous automated operation, Boeing engineers ran tests with contaminated fluid having 50 times more particles than the crash airplane. At no stage did the power control unit jam or reverse, although it required stronger forces to move its input levers, and the testing wore out four hydraulic pumps, possibly because of the high particulate content of the fluid. When the power control unit was dismantled afterward, a fifth of an inch of silt was found inside one cavity, and some of the valves in the servo unit had been pitted and worn by the contaminants. But the unit still worked. Boeing engineers were jubilant and declared that the power control unit design was "tolerant of significantly higher particulate contamination levels than those present in service."

At the second public hearing in November 1995, Boeing hydraulic engineer Richard Kullberg cited his company's tests and played down the need for further research into the effect of silting in hydraulic fluid. First, he said, the system was protected by several filters capable of catching all but the smallest particles. Second, he pointed out that the constant movement of the rudder in flight as commanded by the yaw damper effectively flushed any silt out of the power control unit. The filters removed all particles larger than 25 microns, and the gap between the valves and housing was larger than that. Kullberg reminded Greg Phillips, who questioned him, that the filters had been removed in order to concentrate silt in the power control unit and make the contamination much worse than it usually is. Yet the unit still operated normally.

Paul Knerr followed Kullberg to the witness stand. He was not sure that Boeing's tests proved that contaminates had not been a problem. He predicted that metal particles shed from the valve through wear and tear could build up in sufficient quantities to create some sort of blockage, especially if certain parts of the valves had become worn.

NTSB staffers, members of the Systems Group in particular, were not exactly enthusiastic about Knerr's appearance at the public hearing because he had no direct involvement with any of the groups investigating the crash. But, having indulged the chairman's whim, they believed they could simply pick up from where they had left off. But Jim Hall had a further shock in store.

In January 1996, Hall announced that he was forming a special panel of outside hydraulic advisors to scrutinize the work of his systems investigators to see if they had missed any vital clues or if there were other areas of study they should pursue. The news was a bombshell at the NTSB's headquarters at L'Enfant Plaza in Washington, D.C. Some investigators were angered by what they saw as a slur on their professionalism; others regarded it as the thin end of a wedge that would eventually be used to do away with the NTSB's independence. The fact that Hall had turned to FAA administrator David Hinson for help in composing his panel was seen by some staffers as near to blasphemy, considering the FAA's cozy links with industry and airlines. And there was perplexity when Paul Knerr, a man with no record of air accident investigation, was chosen to chair the new panel.

The objections never erupted into formal protests, at least none that landed on Hall's desk. More than anything else, this reflected the fact that Bernie Loeb supported Hall's plan. During his tenure as director of Aviation Safety at the NTSB, Loeb definitely left his mark on the organization. He had a forceful personality and it was said of him that he took no prisoners. He tended to take a hard line in the drafting of probable cause findings and safety recommendations. If something was unsafe, he didn't mince words in saying so. Both the FAA and plane makers were often targets of his wrath. For example, if the FAA had approved a design or a practice later found to have been a cause or contributory factor of a crash, Loeb wanted the FAA added to the list of culprits. With a reputation larger than life, Loeb pushed his investigators hard. While other heads of departments at the NTSB concentrated on the broader picture and left it to investigators to get on with things, Loeb often got involved in the fine details of a crash probe.

A turning point in his career came when he oversaw the investigation into the sudden detachment of a cargo door on a United Airlines Boeing 747 over the Pacific in February 1989. Nine passengers were blown to their deaths when the floor over the door collapsed and a large section of the airplane's skin peeled away. The pilots succeeded in landing the aircraft without further casualties. Under pressure to produce a timely report, Loeb pushed investigators into writing a probable cause, and they reported that ground handlers had failed to properly close the door before takeoff. Cited as contributory factors were Boeing's failure to produce a fail-safe design and the FAA's failure to ensure that proper door-locking procedures were in place. At the time of the report, the missing door had not been found, but there had been a prior history aboard 747s of door problems caused by damaged locking pins.

Almost a year and a half after the report's publication, the US Navy unexpectedly recovered the missing door. An examination revealed that the manual locking mechanism had worked. The ground handlers had, in fact, properly locked the door. Loeb had to withdraw the original finding and produce a new one, which stated that a short circuit or arcing in wiring connected to the electrical door locking mechanism had probably occurred in flight, causing the door to fly open. It was the first time in the NTSB's history a report had to be withdrawn, and it was an acute embarrassment to all concerned.

According to some observers, this fiasco changed Loeb from a man obsessed with coming to a rapid conclusion about the cause of an accident to one obsessed with ensuring that all the investigative bases had been covered. He wanted the NTSB's decisions to be backed by a mountain of supporting data, regardless of how long it took. At this time, observed one ex-staffer, the style of the NTSB's reports changed. "From the 60s to the mid 80s all the reports, on the most complex of aircraft accidents, were short, to the point with conclusions, probable causes, and recommendations. All of a sudden we were writing encyclopedias. Are we trying to show people how smart we are? No. Are we defending a position? Most likely." When Greg Phillips favored writing a safety recommendation calling for the complete redesign of the systems in the 737's tail less than six months following the crash of Flight 427, Loeb backed the decision to shelve it, arguing that finding out exactly what had happened was the best way of ensuring that the right thing got fixed.

Loeb supported Hall's idea of an independent hydraulics review panel of outside advisors because he believed that bringing in people who could come up with new ideas would be constructive. Admittedly, there were also difficulties. Loeb recalls, "First of all you were bringing in people you didn't know. Were they going to end up talking to the media about the investigation, creating huge problems for us? Were they even going to be as good as we might like them to be? Equally, they would be working with proprietary data, and there are legal implications to that. We got through all those hurdles and I thought it was a good idea to do it, although some others within the staff didn't think so."

Paul Knerr's independent review panel met for the first time at NTSB's L'Enfant Plaza offices in early February 1996. Its members were aviation hydraulics luminaries from the FAA, the US Air Force, NASA, and industry. In addition to Knerr on the industry side, there was Ralph Vick, who had worked at Bendix in the 1960s on the design of a power control unit servo valve for use on the forthcoming Boeing 747. At the very first meeting, Vick recalled how he had to redesign one of his 747 servo valves when it failed a thermal test designed to check the valve's ability to tolerate dramatic extremes between the chilling temperature at cruising altitude and that of hydraulic fluid heated by coursing through sometimes overheated pumps. The valve's moving piston had expanded because of the warm fluid, jamming the body of the unit. The Air Force expert then brought up the crash of a fighter when a power control unit jammed and the wrong flight control surface was deployed. Thermal expansion caused by hot fluid entering a cold unit was blamed. Before long, panel members were asking what if Flight 427's main rudder power control unit had jammed for similar reasons? And, as a consequence, could a stuck rudder have pitched the plane into its death dive?

The notion of thermal expansion of hydraulic valves was not new, and for decades a thermal shock test had been a standard requirement in testing newly designed valves for aircraft. Most servo valves, like the power control unit in the tail, operate in unheated areas outside the aircraft's warm, pressurized interior spaces. At altitude they can become very cold, so in tests a unit is chilled to the sub-zero temperatures it might encounter at, say, 36,000 feet. It is kept at that temperature for a time; then warm hydraulic fluid is quickly pumped in and the valve's input levers are artificially moved to simulate pilot controls. If the unit jams, it fails the test and components are redesigned. If it passes, the unit goes into production. Like the testing of many aircraft components, thermal shock testing is extreme, designed to measure performance in conditions way beyond what might be encountered in the real world. Extreme tests give a comfortable margin for error. But what if that margin was simply not great enough?

Knerr knew the panel needed to look at other aspects of the work of the Systems Group. He was also eager to have the panel consider his pet theory, that silting might have somehow contributed to malfunctioning of the power control unit. But the question of thermal testing had dominated the meeting, and Knerr had to report back to Hall with a proposal that they conduct special tests on the Flight 427 power control unit's ability to withstand thermal shock. Now it was Knerr's turn to be shocked. First, Hall told him, the testing would have to be conducted by the Systems Group, via the party system. And the test criteria could not be imposed from outside; they had to be agreed upon by all members of the group beforehand.

Naively, Knerr had half-expected Hall to give him the green light to take the components away for testing at his or some other lab until he either cleared them or discovered a fault. He had not realized how deeply the party system was ingrained in NTSB operations or how much each of the outside parties was involved in devising and approving tests. In forensic tests like those he now proposed for the Flight 427 power control unit servo valve, he wanted to be able to follow his nose whenever promising leads for inquiry presented themselves. Contamination with silt was a case in point. Side by side with the thermal testing, he wanted to be able to adjust the levels of contamination up and down to see what happened. But the NTSB party system ruled out such a freewheeling approach.

Knerr obviously didn't like the party system:

> ... There were about 12 individuals who come from the airline pilots association, the mechanics' union, Boeing, the primary system company, which was Parker, and USAir, plus people from the NTSB. The tests cannot proceed without all 12 of them agreeing to the direction of the investigation. So every step of the testing, we had to sit down with those 12 individuals and have them approve it. So it takes an awful long time. We would prepare one document which would be the test procedure for the thermal shock and they would send it back saying no, we want it this way and it would go back and forth, back and forth, until everybody agreed and then we would finally do the testing. We couldn't deviate from that testing in any way....

As Knerr saw it, aside from the unwieldy nature of decision making, the party system compromised NTSB's independence. If the participants in the investigation represented the people who designed or built the aircraft and its component parts, or who owned and managed it, or who flew it, the system gave them the opportunity to make sure that whatever kind of testing went on would protect their interests. Obviously, it would be very difficult for investigators to follow their professional instincts and pursue promising lines of investigation. Knerr wrote to Jim Hall that it would be better to just get a group of hydraulic experts and turn them loose to investigate however they see best, with no strings attached.

But Hall's hands were tied by the party system and by the ground rules he had agreed upon for Knerr's committee. NTSB staff and their groups would continue to investigate under the terms of the party system. Reviewers like Knerr would look over their work and approve it, criticize it, or suggest avenues for further exploration. Still, Hall did make one concession to Knerr. The testing could take place at Canyon Engineering's Valencia facility instead of at Parker Hannifin or Boeing. True to form, there followed a lengthy period of consultation between the parties about the shape that the tests would take. Six months would elapse before agreement was reached.

apter 15 HUMAN FACTORS REVISITED

Malcolm Brenner kept thinking about the stories of Captains Gary Higby and Don Widman, which Boeing representative Curt Graeber had brought to the attention of the reconstituted Human Factors Group in early 1995. The captains' first officers, who were piloting the planes, each had a seizure on final approach and stomped hard on a rudder pedal. It could happen again. If the captains had done nothing when their first officers became incapacitated, the 737s they were flying would indeed have rolled until they were inverted, and then they would have crashed.

Then again, there were major differences between these incidents and what Brenner and his group were investigating. First, Flight 427 went out of control at more than 5,000 feet above the ground and could not be recovered. Higby was at 1,500 feet and Widman a perilous 600 feet above the ground when their incidents happened. Yet both captains were able to rescue their aircraft and land safely. Both men had been startled—Higby probably more than Widman when his pilot started to scream—and while Higby admitted to a slight delay in reaction time and Widman failed to notice the rudder command, neither made wrong wheel or rudder adjustments or did anything foolish with the control column. Higby even had the presence of mind to advance the throttle of one of his engines to overcome the yawing and rolling force of the rudder.

"The point ... was that in both cases they land the planes safely, and that's why you have a second pilot up there," Brenner reasoned. "The system worked, that's why you have two pilots."

Brenner's mind kept going back to the March 1991 crash of United Flight 585 at Colorado Springs. His colleagues in the Aircraft Performance Group had uncovered similarities between that and the Flight 427 crash. Brenner had gone the extra mile in digging out the histories of the Flight 585 pilots, Captain Harold Green and his First Officer Trish Eidson. He had listened to the Flight 585 cockpit voice recorder over and over until he knew those pilots so well he could almost predict what they would do in an emergency. He tried to imagine them in the place of Higby and Widman, and he was convinced they would have performed as well. During the attempt to solve Flight 585, there was almost no pressure to blame the crew, as was occurring in Flight 427's investigation. "I always felt in the back of my mind that something unknown must have happened there, but we just couldn't find it," he would say afterward.

Brenner interviewed pilots who had flown with Green and Eidson, and they all spoke highly of their abilities and their approaches to the job. One man who flew several times with Green recalled how, when they were preparing for a difficult approach, Green said he had no hang-ups with aborting the landing and doing a go-around. He displayed no symptoms of the "I'll-get-there-on-time-at-any-cost" syndrome that often leads to accidents, especially in bad weather.

In most incidents, there were human performance issues that would bother Brenner. There were few "100 percent perfect" flights, but he found Green and Eidson were beyond reproach. The cockpit voice recording demonstrated as crisp and clean a crew performance as he had ever seen in his time at the NTSB. They performed every checklist, they gave each other clear briefings, they had good relations, there was a little bit of humor, they discussed the weather a number of times, they discussed the speeds they were going to use, and they added a little extra speed for coming in. Eidson called ahead to get a wind shear report and said, "I'll watch the instruments like it was my mother's last minute." For Brenner, this was a case where crew procedures were working near perfectly, a credit not just to the individuals in the cockpit but to United Airlines, which had trained the pair. Given the impeccable performance, given just a strong feeling that these two pilots were as sharp as any he'd ever listened in on, Brenner had a sense that something out of the ordinary had to have occurred.

What had happened to Green and Eidson bore alarming similarities to the fate that befell Captain Germano and First Officer Emmett in Flight 427. Neither crew had an opportunity to verbalize what was happening. It was as if they didn't know what hit them.

Brenner was monitoring the amazing progress his colleague Jim Cash made in squeezing ever more data and information out of Flight 427's cockpit voice recording. Could Brenner, perhaps with outside help, take the same approach? Was there an opportunity for a voice analysis along the lines he had commissioned for the *Exxon Valdez* tapes? Could there be another interpretation of the grunts and other exclamations from Germano and Emmett that might betray their state of mind and what led to their actions? At the back of his mind were the breakthroughs that Alfred Belan had made in Moscow. Perhaps Belan's techniques could help reveal something new about the events in Flight 427's cockpit during the last 30 seconds of flight.

When Brenner ran his ideas for deeper voice analysis work past his Human Factors Group, it accepted the proposal, though its reaction was lukewarm. Robert Sumwalt was tempted to pigeonhole the proposal with what he considered to be some crazy ideas emanating from Graeber's side of the table. But he eventually supported it as an attempt to uncover new data that might be useful. Graeber himself was mildly disparaging, although he too eventually supported it, as did Carriker. Phyllis Kayten also supported it, despite misgivings. She was a fan of Brenner's work on the *Exxon Valdez* tapes and often used his findings in lectures she gave. But this time she thought Brenner's approach was a bit "fruity," and she had trouble seeing its relevance. She also had doubts about the validity of Russian psychology dating from the time when she sat on an International Civil Aviation Organization committee drafting a handbook on human factors training for air traffic controllers. The editorial committee was unable to use a Russian contribution because it lacked relevance. Now she couldn't see the real relevance of letting the Russians loose on the Flight 427 tapes.

Brenner assembled a separate group to handle the speech analysis, himself plus Jim Cash and David Mayer, another NTSB acoustics expert. They were not the first to use analyses of pilots' speech to probe the cause of a crash. In August 1985, following the crash of a Boeing 747 and the loss of 520 lives, Japanese investigators used speech analysis to discover the level of stress on the pilots and to determine whether they were suffering from hypoxia, or oxygen deprivation, a condition that can cloud judgment and slow response times. In the 747 accident, it turned out, the cone at the rear of the pressurized cabin blew out, destroying the tail fin and the rudder and leaving the aircraft without hydraulics and flight controls. The accident had depressurized the airplane at above 20,000 feet, where the air is extremely thin, but it was more than 10 minutes before the flight crew donned their oxygen masks. Despite the emergency, the crew said very little to each other and took several minutes to respond to radio calls, a sure signal of oxygen starvation. Matters improved slightly when they descended below 20,000 feet, and even more when they finally donned their masks. But their uncontrollable aircraft was destroyed when it struck a mountain.

When two small aircraft on training flights collided and killed all aboard in New South Wales in May 1988, Australian investigators suspected that the air traffic controller, who had put the two aircraft on conflicting flight paths, was overworked and under pressure. Analysis of his tape showed that he was speaking at a rapid rate just before the accident and misarticulating some syllables as he attempted to get through his work quickly.

In another crash, Brenner himself had analyzed the air traffic control tapes of a light aircraft flown by a single pilot who died when it went down in

Texas in August 1991. The speaking rate of the 62-year-old pilot had slowed dramatically between airports, and he appeared tired and confused. Two men who knew him well listened to the tapes and said the pilot sounded almost like “a different person.” Hypoxia was ruled out because he was flying below 10,000 feet. Brenner concluded that the man, who had had a heart condition, probably had a heart attack or stroke. Coronary disease was also indicated in the tapes of a man who died along with his passenger when their Beechcraft suddenly departed level flight and crashed in Montana in August 1995. He displayed breathing difficulties toward the end of the tape.

If anybody could tell whether Germano or Emmett had suffered a seizure or was under intense physical strain, it was Scott Meyer, one of three outside experts Brenner picked to analyze the Flight 427 tapes. Head of Aviation and Operational Medicine at the Naval Aerospace Medical Research Laboratory in Pensacola, Florida, Meyer had done medical research for NASA and was a lead researcher for a congressionally funded project to find out how physically strong a pilot needs to be to become a navy flier. Before that he had investigated the breathing patterns of navy pilots under high G forces and measured the stress that straining to overcome high-G maneuvers placed on their systems.

Meyer concentrated on the rate and depth of breathing of the Flight 427 pilots. Under physical exertion, the rate of breathing increases first, then breathing becomes deeper as more air is inhaled and exhaled to cope with the increasing demand of the body for oxygen. The quality of the tape was excellent and Meyer could hear the pilots breathing perfectly. He even spotted when they yawned during the flight. He logged the captain’s normal breathing at 30 breaths per minute, slightly higher than usual but not abnormal. When the emergency began, Meyer noticed that Germano took one sharp deep breath just before he said “Whoa.” Meyer pictured him being startled by a sudden departure from normal flight. Then Germano’s breathing rate increased dramatically, once reaching just short of 60 breaths a minute, almost a hyperventilation level. If this was indeed a nervous system response to the emergency, Meyer predicted that the captain’s heart rate, body temperature, and blood pressure would also have increased. Yet Germano’s depth of breathing remained the same throughout the tape. This means that the pressure he was under may have been psychological or emotional, but it was not physical. Whatever Germano was doing was not taxing his strength. And his breathing was that of an alert, conscious pilot, not of an unconscious or incapacitated man.

Things were different on First Officer Emmett’s side of the cockpit. He was piloting the airplane and looking out of the cockpit window to locate the

Jetstream commuter aircraft mentioned by the air traffic controller. He had just said, "I see the Jetstream ..." when he interjected "Zuh," followed by two rapid grunts. Meyer closely examined those grunts and determined they were the result of physical exertion, the second more forceful than the first, as though Emmett was trying to "break out," or override, the autopilot. (When force is used on the control wheel or column, it causes the autopilot to disconnect.) In times of physical exertion, the glottis, which produces the voice, is often temporarily closed, causing the act of exhaling to build up pressure inside the body. When the glottis is then relaxed, the released breath is often forceful and accompanied by a grunt, like a weight lifter's. Meyer could not tell for certain what sort of muscles the first officer was using. It could have been his arms controlling the wheel and the column, or his legs on the rudder, or both. And after those first two grunts, he did not appear to be under any physical strain, particularly after the autopilot disengaged. And he, too, was conscious and alert.

At NASA's Ames Research Center in California, a second expert engaged by Brenner, Barbara Kanki, took a different approach to the tapes. While Meyer had concentrated on the men's breathing and nonverbal sounds, she listened to what they actually said or tried to say. A Ph.D. psychologist, she had studied cockpit resource management, examining how crews work together and how they communicate. In her opinion, the Flight 427 crew worked well during the first 30 minutes of the tape, with all procedural actions completed and nothing left hanging. During the emergency phase, however, they said little she could analyze. Obviously, a problem existed, but whether either pilot had identified its nature was unclear.

For example, when the Germano said "Hang on," he could have been speaking literally, in other words asking the first officer to hold on to something because of the motion of the airplane. Or he could have meant, "Wait, I'm thinking," or been referring to the aircraft and waiting for it to respond, or any combination of the foregoing. Kanki had no way of knowing. When the captain then said, "What the hell is this," she did not know to what he was referring. Possibly the first officer knew what he meant because, as Kanki points out, something may have been very evident to both of them, or the captain may have been pointing to something or looking at it. "We have no way of knowing from the words alone whether the captain and first officer are completely in tune with each other and therefore don't need to use referents and complete sentences or whether they are responding to different aspects of the situation," she wrote Brenner in December 1996.

Alfred Belan in Moscow was the last to respond, in late March 1997. He found that the crew of Flight 427 went through all of the classic reactions of a serious emergency, ranging from sudden surprise, through increasing psychological stress, then panic. Captain Germano's first reaction was one of sudden surprise, Belan decided, and this was evident from his use of the words "sheez" and "whoa," and from breathing disruptions. It could have been something he saw or heard, but these were also the types of reactions commonly encountered in sudden, unexpected movement. Over the next few seconds, Belan's analysis of the captain's breathing indicated heightened psychological stress. His breathing became rapid and his speech became louder and faster. But there was no grunting or forced inhaling and exhaling that might indicate physical efforts in using the controls. His actions at this stage were limited to issuing commands ("Hang on") and trying to understand the situation. His brief statements conveyed little or no information, a further sign of psychological stress. Belan's interpretation of the captain's use of the phrase "Hang on" was that he was trying to understand what was happening to them, and his subsequent utterance, "What the hell is this," suggested that he did not know.

Thirteen seconds from the onset of the emergency, Captain Germano entered the second level of emotional stress, according to Belan. The frequency and intensity of his voice changed further and he was becoming emotional, as shown by his use of the phrase "oh God." He was still collected enough to answer the call from air traffic control. "However, his answer was incomplete and it is obvious that the situation was unclear for him," Belan reported to Brenner. Just over 20 seconds into the emergency, Germano was yelling "Pull" and had entered into the highest stage of emotional stress, panic. He could not act in accordance with the situation. His speech had reached a high intensity and frequency, and he was issuing a command ("Pull") that was, at the very least, inadequate. Finally, he started to scream. By this stage, Belan reckoned that the captain had started to participate in attempting to control the airplane. Short inhalations of breath immediately follow each utterance of "Pull." These are characteristic of high physical loads, such as pulling back on the control column, although Belan conceded that they could be caused by attitude (such as being upside down) or by high G forces.

Emmett's first reaction was also one of surprise in Belan's analysis. "The word 'zuh' has no meaning, increasing the likelihood that it was an involuntary exclamation due to surprise rather than an intended statement," Belan said. Soon Emmett's breathing included grunting and forced exhalations that indicate he was actively in control of the airplane. However, normal use of cockpit controls would not produce this type of reaction, suggested Belan. "These sounds suggest that the first officer was struggling unusually hard, for example if he was pushing a control against its stops or if he was experiencing an unusual resistance in the use of a control." At one point, the pressure appeared to have relaxed, because when Emmett said the words "oh shit" some 7 seconds into the emergency, there was no evidence of grunting or straining or forced exhalation in his speech. But when he next says "oh shit" about 21 seconds into the emergency, there is constricted breathing and the data from the flight data recorder confirms that at this stage the control column was pulled back hard against the stop.

The only word spoken by the first officer in both the emergency and pre-emergency phases of the flight and which could be validly compared by spectral analysis was the expletive "shit." Emmett had referred to a piece of equipment, presumably some aspect of the flight system he was attempting to program, as "you piece of shit" during an earlier phase of the flight. He spoke the same word again in altogether different circumstances as the airplane tumbled through the sky, and Belan was able to compare the different utterances on the basis of frequency, intensity, and pitch.

Finally, just before impact, Emmett's speech degraded into short exclamations and expletives, indicating high levels of psychological stress culminating in panic. The radio switch on the control column was pressed several times in this period, which Belan suggested could have been Emmett attempting to brace his hands better on the control column as he tried to pull the plane back into control.

The group members had given a lukewarm reception to the speech analysis proposal, but generally had to admit that they were more than just satisfied with the work that had been done. Robert Sumwalt was pleased that the analysis dispatched some of what he considered to be outlandish notions about what the crew should have done, or not done, in the cockpit. There remained a likelihood that they would still find the crew at fault to some degree, but within a narrower band of possibilities. Kanki, Meyer, and Belan's contributions confirmed his earlier impression from the cockpit tapes that whatever had happened had been rapid and unusual.

Phyllis Kayten was pleased by Belan's contribution. "I was surprised that it was much better than I thought it was going to be, it wasn't totally off the wall," she said afterward. "I hadn't expected it to be worth anything. I first thought it was a waste of money, but it was surprisingly legitimate."

Far from being "fruity," as Kayten first suspected, Brenner's approach had actually helped refine their knowledge of what went on in the cockpit. They might not be any closer to knowing what caused the crash, but they were now able to soft-pedal some theories. For example, the suggestion that Emmett had panicked and applied the wrong controls, and that the Germano had been fighting against him, was effectively disproved. Both Meyer and Belan convincingly demonstrated that Germano did not appear to have been applying control forces, particularly in the early stages of the emergency, although he may have been pulling on the column at the very end. Emmett was a larger, fitter man, while the captain had undergone back surgery just a few months earlier and could be predicted to lose a strength contest. Meyer and Belan also poured cold water on the suggestion that the pilots had panicked immediately and used the wrong control, such as inappropriate rudder. The conclusion was that the first officer had continued to attempt to fly the airplane and the captain, far from attempting to wrest control, appears to have been trying to work out what happened.

But the speech analysis findings raised nearly as many questions as they had answered. Rudder movement was a prime suspect in the cause of the crash. If Emmett did not put in the rudder, who—or what—did? And what was causing Emmett to exert himself to the point of making obvious grunting and straining noises when nothing in the cockpit, even the control column, should have demanded that much effort?

Curt Graeber had an idea he was pushing as the cause of the crash. In October 1996, in partnership with his Boeing colleague Mike Carriker, he came up with two research papers for the Human Factors Group to review. One was by Richard Schmidt of the University of California, the other by Wolfgang Reinhart of the National Highway Traffic Safety Administration. They were interesting pieces of work, about automobile accidents in which people pressed the gas pedal instead of the brake and crashed, a process christened "unintended acceleration" (UA) by psychologists dealing with automobile drivers. In a letter introducing the papers, Graeber and Carriker said, "The research findings on UA (although a different venue) are seen as directly applicable to understanding the [Flight 427's] crew's unintended or inappropriate rudder input following the wake encounter." Although everybody by now agreed that the wake vortices of the preceding Delta 727 could not have caused Flight 427 to crash, investigators were still faced with the fact that the accident sequence started just as it reached a position where it could have encountered the 727's wake.

Brenner passed around copies of the papers to other members of the group at an all-hands meeting on October 31, 1996, in Pittsburgh, called to review progress in the investigation. Robert Sumwalt read his copy over breakfast at the Holiday Inn where the conference was taking place. When he finished reading, he was so angry that he went over to Herb LeGrow, the Air Line Pilots Association's coordinator for the Flight 427 accident, and told him he would not be attending that morning's session. He had a letter to write to Brenner.

He angrily pounded out the paragraphs on his laptop. Though he wasn't a Ph.D., he could spot psychobabble a mile off. Graeber and Carriker's submission was a prime example of attempting to shoehorn extremely interesting findings—but findings irrelevant to the case at hand—into the Flight 427 scenario.

"First, we must comment on Boeing's statement that this information can be 'seen as directly applicable to understanding the *crew's unintended* or *inappropriate rudder input ...*'," Sumwalt wrote. "As you are aware, this investigation is still ongoing. No determination of cause has been issued, and there is no factual information to support a statement that the crew was causal to this mishap. Therefore, I request that you disregard this statement. Also, to prevent future blunders of this type, I encourage you to advise Messrs. Graeber and Carriker of the status of this investigation so that they can keep their remarks in perspective."

Point by point, Sumwalt proceeded to criticize Graeber and Carriker's interpretation of the research papers. First was the issue of where most UA cases occurred, and that was when drivers had just started the car. They put their foot on the gas pedal by mistake, selected Drive or Reverse, and shot forward or backward and crashed. The incidents did not occur when the car was cruising along a highway. There was strong evidence cited in the papers that most UA accidents occurred because of a driver's unfamiliarity with the controls. Many involved rental cars, parking lot attendants, or new owners. "You will recall the flight crew of USAir 427 had literally thousands of flight hours in this exact type of aircraft. Unfamiliarity was not a factor. Further, they had been seated in the accident aircraft for several hours that day, including at least the final 30 minutes of flight." In any event, Sumwalt added, putting the wrong foot on the incorrect rudder pedal in a 737 cockpit is almost physically impossible due to a structural divider between the two pedals.

Next, Sumwalt noted that in laboratory simulations, "drivers" who pressed the wrong pedal immediately noticed their error and corrected it, implying there was no reason an airline pilot would not do the same. Most of the drivers involved in UA crashes were over 60 years of age and the rate of their involvement increased dramatically with age. "To keep this in perspective, the captain of USAir 427 was 45 and the first officer was 38 years old," Sumwalt remarked. Finally, Sumwalt had also noticed from the literature that slightly more women than men were involved in UA accidents, and the accidents also tended to involve people of less than average height. "As you know, both pilots of USAir 427 were men, and both were considered to be relatively tall."

Sumwalt sat back in his hotel room and surveyed his handiwork. He was tempted to carry on, but after writing 800 words across two tightly spaced pages, he decided he had sufficiently demolished Graeber and Carriker's argument. He had the hotel print a dozen or so copies of his remarks, then he marched into the conference room where he distributed them to the members of the Human Factors Group, as well as to Hauteur, LeGrow, and George Snyder, the new USAir accident coordinator.

Then he took his seat at the table, satisfied that he had done a good morning's work. He was coming to realize that having a Ph.D. might not be such a big deal.

Part Three

517 AND AFTER

pter 16 THE EASTWIND INCIDENT

John Cox was not impressed by Paul Knerr—the kind of "guy with a theory" that crash investigators tend to loathe. As a member of the Systems Group and the lead ALPA representative, Cox subscribed to the party system of crash investigation. You did things by the rules and with the prior agreement of everybody else. There were no solo missions, no people operating on their own. At this stage, Cox felt his allegiance leaning more toward the NTSB than his union, which of course appointed him to the investigation. He liked the system "because it tends to prevent anybody trying to whitewash anything. It's a bit adversarial but it gains a lot of strength from this. Nothing gets through without a good look by a lot of experts, and if there is any question about it, the system gleans the truth pretty quickly."

Paul Knerr found all this out the hard way, by seeing his wide-ranging proposals for cold-soak testing—thermal shock testing—of the 737 rudder power control unit whittled back as both Boeing and the NTSB converged on a formula that would satisfy both sides. Knerr had had different ideas. He wanted to be able to follow his nose when intriguing possibilities presented themselves. If a valve started to stick, for example, he wanted to be able to remove or add contaminants to the hydraulic fluid to see if the stickiness eased or the valve jammed totally. That way he would have a clue as to whether contamination was, or was not, a significant factor in making a valve misbehave.

But the Systems Group, after endless weeks of deliberation—some of it quite heated—had decided that the hydraulic fluid quality must approximate what was recovered from cavities in Flight 427's PCU—no more, and no less. They eventually settled on Class 10 contamination fluid drawn from airplanes in service. To some degree, they were taking comfort from Boeing's January 1995 tests, where grossly excessive silting wore out pumps but failed to jam a servo valve. Those tests had not been conducted as part of the party system but were nonetheless convincing.

There were also arguments within the group about the scope of the investigation. Should it engage in blue-sky science, probing any and all ways the

PCU could make trouble? Or should it concentrate on finding out what, if anything, went wrong with the PCU on Flight 427?

Knerr eventually knuckled under to the unyielding stance of the Systems Group, and a series of tests was agreed upon. In the first group of tests, Flight 427's main rudder PCU would be placed in a refrigerated chamber and chilled to the likely temperature it would reach after spending time at cruising altitude. Then, in a variety of ways, it would be brought up to the temperatures it would be likely to encounter as its altitude decreased on the approach to Pittsburgh. At specified intervals, the PCU's input arms would be operated to replicate rudder pedal or yaw damper commands, and the results carefully monitored. Some hydraulic system failures would be caused deliberately. Only then would the more extreme tests suggested by Knerr's review panel be conducted.

During the debates on the tests, questions had arisen about the worst-case scenario for the difference in temperature between the chilled body of the PCU in the unheated tail cavity and the temperature of the hot hydraulic fluid that would be injected into it. This differential, known as Delta T, proved to be a contentious issue.

Knerr's review panel favored a very high number for Delta T and presented some "what ifs" to bolster their worst-case scenario: What if the yaw damper had ceased operating or was switched off, and no fluid was circulating through the chilled PCU? And what if the fluid was warmer than normal? What if very hot fluid suddenly gushed into a super-chilled PCU? Metals contract as they get colder and expand as they get warmer. Could the resulting thermal changes create havoc in the PCU's innards? Although there was no evidence to support a scenario that the yaw damper on Flight 427 was inoperative or switched off, such a thing was possible. And hotter-than-normal fluid could arise from a faulty, overheated hydraulic pump. Such things were not unusual.

The whole testing process worried Boeing engineers, and they took a very tough stance on the procedures they wanted adopted. In the 1960s, this PCU design had passed a rigorous cold-soak test as part of its certification process. It had been subjected to a Delta T far in excess of what it would ever encounter in service. Leery of subjecting the PCU to unrealistically harsh conditions, Boeing engineers came up with a test protocol that they said would prove whether or not it could cope with thermal shock in the worst conditions it was likely to encounter in service. They wanted those tests done first, before moving on to the more extreme tests devised by Knerr's review panel, tests that Boeing said had no relevance to normal operation. Their rationale was that, no matter what the results of using the more extreme Delta T numbers, their tests should prove that the unit functioned perfectly in the real world.

Boeing's engineers supported a worst-case Delta T of 187°F, based on hot hydraulic fluid of 160°F and a PCU chilled to –27°F, the temperature of the interior of the tail fin cavity at 35,000 feet. (Outside temperatures would be even colder, to –65°F, but the friction of air passing over the skin of the tail fin raises the temperature inside.) Before it could be introduced into the PCU, Boeing wanted the warm fluid to pass through a 15-foot coil of cooled pipe to simulate the distance it must travel in the unheated tail fin before reaching the PCU. The friction of the valves at various temperatures should also be measured beforehand. Then, in the event of the unit getting harder to operate, they would know whether or not it was due to a jam, or just getting sluggish due to the increased viscosity of the fluid or other factors. According to Boeing engineers, friction might quadruple at low temperatures. They also argued that in normal everyday service, Delta T could be even lower because hydraulic fluid, which is warmed from passing through heated parts of the airplane or through pumps under pressure, is constantly flushing through the PCU and would tend to keep it at a higher temperature than its surroundings.

The debate swung back and forth throughout the summer of 1996, until a consensus was finally reached and a schedule of tests agreed.

John Cox's jaw dropped when he first arrived at Knerr's company, Canyon Engineering Products, in Valencia, California, in late August 1996. He arrived there with the Systems Group straight from a visit to Parker Hannifin, where the Flight 427 PCU had been reassembled and brought up to service condition by Parker engineers. By comparison to Parker's elegant corporate headquarters, Canyon appeared to be housed in a collection of sheds on an industrial back lot. The lab where the tests were to take place looked to Cox "like the back of a garage." There were benches crowded with an assortment of tools and hydraulic equipment, and Cox's untrained eye had trouble making sense of it all. Since he was a pilot, not a hydraulics engineer, he decided to keep his doubts to himself. But then he noted the pursed lips and furrowed brows of his Systems Group colleagues as they gathered around the equipment that Knerr's people were assembling for the tests.

Canyon did not possess, in its regular array of testing equipment, a chilling unit capable of producing temperatures as low as some of the tests called for. So the company had cobbled one together from a foam cooler box, a slightly more sophisticated version of something you might take on a family picnic, fed by a cylinder of liquid nitrogen through a perforated hose. A new PCU was set up on the bench and covered by the cooler box, which had cutouts to admit the nitrogen hose and hydraulic pipes connected to pumps. Temperature probes and pressure and strain gauges were attached to various parts of the PCU for use by both Canyon and Boeing personnel. Canyon had wired the sensors directly to a computer; Boeing was producing a strip chart printout.

The first tests were conducted on a new PCU prior to testing the Flight 427 unit. Once the hydraulics were pressurized and normal temperature hydraulic fluid started to flow through the chilled PCU, the chilling equipment proved inadequate to keep ahead of the warming effect of the fluid. The PCU's temperature rose rapidly, in one case from –31°F to plus 32°F in just 8 minutes. The test rig was then modified to further cool the hydraulic fluid by immersing some of the tubing in a bath of dry ice and alcohol. Unhappy with this turn of events, the Boeing representatives wanted those tests abandoned until equipment capable of running the original test program could be found. When a majority of the Systems Group voted to go ahead, the Boeing representatives said they would continue only under protest.

The new PCU performed normally throughout all of the tests, even when fluid at 170°F was injected directly into it, bypassing the coils of tubing designed to represent the pipe running through the tail fin cavity.

The PCU from Flight 427 behaved normally until the group performed the extreme thermal shock test and a silting test. After being left to stand for 72 minutes with no yaw damper movement and with all the hydraulic filters removed (in the silting test), it took four times the normal force of about 4 pounds to move the input arm of the PCU.

On the afternoon of August 28, 1996, the Flight 427 PCU was submitted to the extreme thermal shock test. Warmed hydraulic fluid at 170°F was injected directly into the unit, bypassing the cooling effect of the coiled tubing. After 11 minutes, on the fourth cycle in which the input arm replicated rudder pedal movements, the PCU appeared to jam in the full left rudder position. The load on the input arm shot up to 40 pounds for 5 seconds, then dropped rapidly to zero. It should have been no more than 12 pounds. Greg Phillips ordered an immediate halt to the testing to discuss what had just happened. The Boeing strip charts were eagerly, if fruitlessly, scanned for clues by the startled researchers, but there was consternation when an attempt was made to call up the data on the Canyon computer. The operator had failed to save the file.

There were two possible reasons for the input arm jamming. One, the pneumatic device used to operate the arm—a laboratory device not found on an airplane—had itself somehow jammed; two, the PCU servo valve had jammed. When the technicians repeated the test, the same thing happened, but this time it was preceded by a sort of a warning. The first cycle of the input arm was normal, but the next two cycles were more sluggish. Then the input arm appeared to jam in the full left rudder position on the fourth cycle. This time the load shot up to 120 pounds for about a second. Then it dropped to zero and the unit resumed normal operation. That evening, Phillips declared a halt to the testing and ordered the PCUs carefully packed up for hand-carrying back to Parker Hannifin for inspection. The examination showed that neither the new PCU nor the Flight 427 PCU appeared any worse for wear after the tests. To almost everyone's frustration, the Flight 427 unit revealed no clues as to what had happened during testing. There were no scratches or marks of any description on the surfaces of the valves or on the inside surface of the unit. Whatever circumstances had caused the interruption to the cycling of the input arm had vanished as inexplicably as it had arrived.

Canyon's failure to capture the critical test data, as well as the inconclusive results of the testing at their facility, helped Boeing successfully argue that the testing should be repeated using superior equipment in its Renton plant. Boeing engineers hinted, for example, that during the failed tests, ice might have formed on some of the complex input linkages on the outside of the Flight 427 PCU and remained unnoticed because of the lack of an inspection window on the makeshift Canyon chilling system.

Three days before resumption of the Systems Group testing, which was scheduled for October 7, 1996, at Boeing, a specially instrumented 737 took to the air to measure the temperatures of the external air, the air inside the tail fin, the PCU, and the hydraulic fluid. The flight revealed that the temperature of the PCU body at 30,000 feet was a relatively warm 20°F, but below freezing nonetheless, while the outside air temperature was –40°F.

As it turned out, in the more sophisticated test environment at Boeing, the results were almost an exact repeat of what had happened at Canyon. The new PCU worked perfectly throughout all the tests, but the Flight 427 PCU jammed during the extreme thermal shock tests. The group performed a test to see what happens to supercooled hydraulic fluid. At temperatures down to –110°F, it became thicker and a light coating of ice crystals was visible on the surface, but it still poured. At a colder temperature, –154°F, the fluid no longer poured and had a gelatinous consistency.

In late November, the Systems Group returned to Parker Hannifin for a further examination of the Flight 427 PCU. Once again they used a new unit for comparison, but this time they had a third unit to test. It was from an airplane that, had circumstances been different, might have met the same fate as Flight 427.

Some six months before, at about 10:00 P.M. on June 9, 1996, Brian Bishop, an Eastwind Airlines captain, was descending into Richmond, Virginia. He was flying a Boeing 737-200, Flight 517 from Trenton, New Jersey, with 55 passengers and crew aboard. Bishop was hand-flying the airplane with the autopilot switched off, as was his habit on approach. As the airplane passed through 5,000 feet on its descent into Richmond, he felt a few brief knocks on the rudder pedals, although they did not move. He glanced across at his co-pilot, but he had his feet on the floor, not on the rudder pedals. Bishop, on the other hand, tended to rest his feet lightly on the pedals. Suddenly, as the airplane descended through 4,000 feet, it yawed abruptly to the right and started to roll in the same direction. Bishop pressed hard on the left rudder pedal and also swung the control wheel to apply the left aileron. Alarmed, the co-pilot glanced over and was amazed to see his captain wrestling with the controls and practically standing on the left rudder pedal. Bishop found that the rudder pedal was stiff and did not respond as he expected. The airplane continued to roll right. He reached over to the throttle levers on the center console and advanced the right engine throttle. As the right engine powered up, it helped swing the airplane back toward level flight, but then the roll to the right returned. The two pilots then went through an emergency checklist, which included switching off the yaw damper, and the airplane returned to level flight shortly afterward.

Like many pilots who have a tendency to either exaggerate or underestimate the effect of an in-flight emergency, Bishop was wrong on several counts according to the flight data recorder. He initially said the airplane rolled right up to 30°, but the flight data recorder showed that 10° was the maximum right roll. On the other hand, he said the airplane might have rolled "a little" to the left as a result of the combined commands of rudder, aileron, and differential engine thrust, and the flight data recorder showed it rolled up to 15° left at one stage. Bishop thought he had commanded about 45° of wheel, but NTSB calculations suggested it was closer to 60°. However, it was beyond doubt that something unexpected and extremely dangerous had happened, given how close the plane was to the ground, and that the pilots had wrestled the plane to bring it back under control.

This was not the first time that this particular airplane had given Captain Bishop a fright. The previous month he had just taken off when a series of unexpected taps he felt on the rudder pedals caused him to return to the airport. Mechanics replaced the main rudder power control unit, and the airplane was back in service the following day. On June 1, another pilot was surprised when the same airplane had two slight yaws to the left about 30 seconds apart, although no movement was felt on the rudder pedals. On June 8, a third pilot said the airplane rolled right when the yaw damper was switched off. That was enough for the maintenance department, which pulled the airplane into the shop overnight and replaced several yaw damper components. The following morning, June 9, 1996, Captain Bishop and his co-pilot took the airplane up for a test flight. It behaved impeccably and they signed it back into service. That night the Richmond incident occurred.

When NTSB investigators examined Flight 517's rudder system, they discovered a few curious things. First, the yaw damper had been rigged incorrectly. Instead of sweeping the rudder no more than 3° in each direction, it could go only 1.5° left, but 4.5° right—50 percent more than it should. But that didn't explain where the apparently uncommanded rudder movements came from. Another clue was found in chafed yaw damper wiring, where a short circuit may have occurred, and in moisture penetration that may have also contributed to an electrical fault. (Later, the PCU and the yaw damper coupler, which connects to the main rudder PCU, were removed and replaced, and no more problems were encountered.)

When the Flight 517 PCU was stripped down and compared to a new one and to the unit from Flight 427, it emerged that each required a different rate of input to reverse when jammed. Why, was not immediately clear. It also emerged that the clearance between the secondary valve and the body of the PCU varied greatly between the units. The new unit had the greatest clearance, Flight 427 the least, with the Flight 517 unit somewhere in between.

The most useful information to emerge from the Flight 517 incident was the experience of its pilot. Here was someone who could describe what had happened in an emergency that bore at least some resemblance to what had happened to Flights 427 and 585. Bishop had gone through something that appeared to be a rudder hardover and lived to tell the tale. NTSB personnel and Boeing experts rushed to Richmond, where the airplane had been quarantined. Boeing fitted the plane with PADS, a portable but extremely sophisticated flight data recording system that logged far more than the paltry handful of parameters recorded on the Flight 517 recorder, and Bishop was invited along for a flight test to replicate his experience. The yaw damper system was also rigged to perform a hardover by a command from one of the FAA observers on board.

The tests were marred by rancor. Tom Hauteur accused Michael Hewett, a Boeing test pilot in the Operations Group, of badgering Bishop during a debriefing session and trying to impose his own interpretation of what had happened over Richmond. John Cox had hoped to participate as a member of both the Operations and Systems groups, but Eastwind was rabidly anti-union. Because Cox worked for the Air Line Pilots Association, he was barred from being on the plane during the test flight. It took much persuasion to allow him even to sit in on the official debriefings of Bishop. Eastwind management, despite having come within a hairsbreadth of an aviation disaster, insisted that Cox act as an observer only, and not ask any questions.

Cox did witness the exchanges between Hewett and Bishop, but being more or less forced to accept his observer-only status, he did not intervene. He felt that Hewett was being arrogant and was leaning on Bishop, but Bishop was no pushover and held his own.

Hewett was proposing that Bishop had experienced a yaw damper hardover, a relatively minor incident and easily handled with opposite wheel. After reviewing the flight data recorder, Boeing and the NTSB had agreed on an estimate of about 7° for the rudder movement that Bishop might have experienced. In the subsequent test flight in the same Eastwind airplane, this rudder command was intentionally triggered while Bishop, not expecting the command, was at the controls. Bishop reacted expertly, using the control wheel and rudder to right the airplane. The force required on the rudder pedal was only a fraction of what he encountered on the night of the "incident," he said. The test flight case was really very different from the real emergency.

When Jim Hall was briefed on Flight 517, he was furious. Here was an incident that might have helped unlock the secrets of the 737 rudder, maybe even help solve the crashes of Flights 585 and 427. Yet, after taking the airplane out of service for almost three weeks, interviewing its pilots, analyzing some of its parts, and flight-testing it, they were no closer to a solution. This made him all the more adamant about having flight data recorders that kept track of more parameters. On June 28, 1996, he wrote to FAA chief David Hinson, complaining at the lack of progress in ordering that improved flight data recorders be retrofitted to US aircraft. For the first time, he officially drew parallels between 585, 427, and, now, Flight 517. "Under slightly different circumstances, the Eastwind incident could have become the third fatal B-737 upset accident for which there was inadequate flight data recorder information to determine the cause," he wrote.

The purpose of the letter was, of course, to underscore Hall's frustration and impatience about the data recorder situation. But perhaps more startling was the explicit statement, from someone at the highest level of the NTSB, of a problem lurking in the tail of the most widely flown commercial jetliner in the world.

And it wasn't just Hall who felt that they were dealing with something more than three discrete, unconnected events. After the Flight 517 incident, Tom Hauteur had concluded that he was not simply investigating the Flight 427 disaster. Instead, he finally came out and said to himself, he was probing the entire Boeing 737 fleet. It was something he had known in some way, of course, ever since he'd been delegated to assist Panamanian authorities investigating the Copa Airlines crash in 1992. The Copa 737's dive was similar to the path traced by Flight 585 in Colorado Springs in 1991, and it was tempting to look for a connection between the two. But faulty cockpit instruments caused the Copa crash, and he found nothing in it that seemed to have any connection to Flight 585. Indeed, a connection could have meant there was something about Flight 585 that the earlier investigation had missed.

Boeing, the world's greatest plane maker, had staked an important part of its reputation on attributing the crash of Flight 585 to a rotor, a mysterious bundle of whirling air. Everyone agreed that the strange weather in Colorado Springs was worth considering, but dozens of other aircraft, some heavier, some much lighter (and more vulnerable) than Flight 585, had come and gone from the same wind-tossed airfield that day without any mishap. Why was Flight 585 so unlucky? Now, with the benefit of hindsight and some further work by the National Council for Atmospheric Research, Hauteur realized that the rotor theory was increasingly unlikely.

Flight 585 had triggered something in the airline community, and a small but steady trickle of reports about 737 pilots momentarily losing control of their aircraft came into the NTSB. Systems expert Greg Phillips became the repository of that growing body of knowledge. Excited with each new report, he was doomed to be disappointed as time after time the problem turned out to be caused by misbehaving yaw dampers. Designed to keep an inherently unstable jet plane on the straight and narrow, a yaw damper pushes the rudder no more than 3° to the left or right. Flipping suddenly to that extent is enough to spill drinks or badly bruise rear flight attendants if they lose their footing. But any skilled pilot would be expected to bring the plane under control by judicious use of the wheel to counter the rudder movement with the aileron. Then quickly running through the checklist, the pilot would turn off the yaw damper and the problem would go away.

After the 1994 crash of Flight 427, the trickle of rudder incident reports turned into a stream. By the time of the 1996 Flight 517 incident, the list had grown to 90 items from all over the world, with 86 of them dating from after the 1991 crash of Flight 585 at Colorado Springs. (Subsequent investigation absolved the rudder from blame in some of these incidents.) Hauteur had long suspected that the 737 incidents being reported were so commonplace that some pilots hadn't bothered to make a serious issue of them. But there were also a few that could not be explained by the yaw damper. Was there an intermittent rogue element in the 737 tail capable of producing effects worse than the yaw damper, or were pilots doing silly things with their rudders? And under just what circumstances was a 737 pilot supposed to use the rudder, anyway?

One might think such a fundamental question would have if not an easy answer, then at least a definitive answer, or a set of definitive answers. But this was not the case. In fact, the subject of the rudder produced some curious reactions from Boeing. In an early 1995 communication, the company held that the Flight 427 pilots had seriously erred if they used the rudder to counteract the roll arising from their encounter with the wake vortices from Delta Flight 1083. The allegation provoked a stormy reaction from the Air Line Pilots Association, which held that if a control was on the airplane, it was there to be used. Later, the NTSB had Malcolm Brenner write up a series of questions for Boeing, designed to elicit the plane maker's official stance on the use of the 737 rudder, so that a line could be drawn between proper and improper applications.

To Brenner's questions about where pilots should place their feet, the use of the rudder, wake turbulence encounters, and recovering from excessive roll, Boeing answers were remarkably unspecific, not to say noncommittal, and often declared little more than the fact that the company expected pilots to have been previously trained in these matters:

> BRENNER'S QUESTION: Where does a pilot place his feet during the cruise, and during descent?
>
> BOEING'S ANSWER: We feel that for a student currently entering transition training today, feet and hand placement are compulsory training learned early in basic flight training. Since these are basic to airplane operation these points are not specifically revisited in Boeing training nor do we publish any written guidance on this subject. Pilots are expected to have feet and hands placed to ensure the desired flight path is maintained in any stage or phase of flight.
>
> BRENNER'S QUESTION: What is the recommended procedure for how the rudder is used during normal approach and landing?
>
> BOEING'S ANSWER: As stated in the response to the first question, this is considered one of the very basic aspects of flying for which Boeing does not provide specific training or instruction. In Boeing airplanes, use of rudder for coordinating normal flight maneuvers is not necessary. As in any airplane, rudder is used for an engine failure

> condition, in performing proper crosswind landing techniques and for directional control after landing.
>
> BRENNER'S QUESTION: Which is preferred, handflying or using the autopilot in wake turbulence encounters?
>
> BOEING'S ANSWER: We expect that flight crews have learned through past training and experience how to maintain the desired flight path during foreseeable flight conditions, which include wake turbulence encounters.
>
> BRENNER'S QUESTION: What is recommended recovery from excessive/severe roll? What is proper rudder input, power setting, control column position?
>
> BOEING'S ANSWER: We assume that flight crews have previously received training on how to avoid and recover from excessive or severe roll conditions. As this training is considered prerequisite to airplane operations, Boeing does not provide written guidance or a published procedure for these conditions.

The company was more forthcoming on actions to be taken at the onset of a stall and on the issue of one pilot handing over control of the airplane to the other. On the latter, the company advised against transferring control during a recovery procedure. But apart from pointing out that the rudder could be used to balance the airplane after one engine failed, when landing in a crosswind, or to keep the plane straight on its post-landing roll, Boeing was silent on other uses for the 737 rudder, refusing to dictate when it should or should not be used. USAir, on the other hand, was much more definite in its replies to the same questions. It said rudder pedals were to be used "at any time aircraft yaw correction is required." Rudder pedals could also be used in responding to excessive roll "to counteract adverse yaw." Why Boeing was not prepared to officially state the obvious was far from clear.

Tom Hauteur would have given his right arm for hard digital information on exactly what controls the pilot was employing on Flight 585, Flight 427, Flight 517, or any of the growing list of incidents where an unusual rudder movement was suspected of causing a problem. If Flight 427's flight data recorder had been fully instrumented, and the data showed a pilot putting his foot down hard on the wrong rudder pedal and keeping it there, as Boeing was now postulating, then the investigation might have been wrapped up in a few months with a verdict of pilot error.

Boeing's answers to Malcolm Brenner's questions were lacking in substance. At the same time, Boeing was strongly suggesting that pilots, when startled, often do the wrong thing with their controls. They put in left rudder when right is called for, and nowhere, said Boeing, was there a more glaring example than the Sahara Airlines crash at New Delhi in March 1994. It appeared that a trainee pilot had pushed the wrong rudder pedal when his instructor simulated an engine failure during a takeoff exercise, and the aircraft nosed over and crashed, killing all aboard as well as several people on the ground.

What exactly was right or wrong rudder use, Hauteur had begun to wonder in 1995. A baseline study of exactly how pilots use their rudders when flying 737s was needed. Fortunately, there was a way to do this. A significant portion of the European fleet carried flight data recorders with many more parameters than their US counterparts. Several European airlines had fitted out their planes with a Quick Access Recorder, or QAR, from which data could be routinely downloaded into a computer database. It drew information from the same unit that drove the main flight data recorder, but unlike the main recorder, it was not crashproof.

British Airways had developed the use of QARs to a fine art. Their data storage tapes or disks were routinely taken off the aircraft and the data put into the airline's comprehensive database, which, in turn, was monitored to study how their aircraft were being flown. Company officials could soon spot a pilot who repeatedly made hard landings, or banked excessively, or used too much (or too little) power during a climb. In many cases, British Airways' QARs measured even more parameters than what was mandated, and sometimes at a higher rate—say, several times per second as opposed to once per second. For its fleet of Boeing 737-400 aircraft, for example, some 200 parameters were routinely monitored. While flight data recorders on newer US aircraft were set up to measure either rudder pedal or actual rudder position, none did both. But in the British Airways 737-400 fleet, both parameters were captured.

Tom Jacky recruited Anne Evans, a senior investigator specializing in flight data recorders at the UK's Air Accidents Investigation Branch, to help set up a research program into rudder use aboard British Airways 737s. A firm of consultants, Flight Data Services Limited, was employed to analyze the data. Additional software had to be written to enable the British Airways computer to recognize and log the data wanted by the NTSB, which now included wheel use. All this, of course, cost money, and Hauteur tried to spread the extra $12,000 cost for the new software by having the appropriate parties chip in. The Air Line Pilots Association refused to pay its share, saying its accident investigation budget was already seriously extended because of Flight 427.

The rudder-use data analysis program ran from October 1995 until March 1996 and covered some 57,000 flying hours spread over 27 airplanes. Investigators were able to segregate yaw damper-commanded rudder movements from pilots' pedal-commanded movements. The data showed that most rudder movements were miniscule, and more than 97 percent were less than 1.5°. Only a handful went beyond that to the yaw damper range of plus or minus 3°. Wheel movement was equally undramatic: in 99.9 percent of the samples it was 20° or less. In one sense, the British Airways research was disappointing because it failed to reveal any serious anomalies in aircraft behavior, thus providing no clear culprit. No Flight 517-style incidents occurred. In another sense, the data were reassuring because they showed that in normal use the 737 did nothing to startle pilots. It was an amazingly steady airplane. The data also failed to demonstrate any tendency among pilots to do silly things with their airplanes. There was not a single example, in the almost 60,000 hours of monitored flight time, of a pilot putting in anything even close to full rudder.

If the pilots weren't responsible and the equipment didn't malfunction, what, then, had happened to Flight 427? Various bizarre scenarios were discussed at meetings and tested. One of these involved a very fat man falling through the floor and landing on a rudder cable below; since he was unable to get out of the hole he had created, the pressure of his immense weight on the cable was enough to pull the rudder over hard to the left, plunging the airplane into its death throes.

This scenario came up at a Systems Group meeting debating the possible significance of a Maintenance Group report that a portion of the cabin floor on Flight 427 had a weak, spongy section, the consequence of a temporary repair. The "fat man" hypothesis was actually tested at Boeing in late February and early March 1996 on a retired 737-200 that USAir was donating to the Museum of Flight in Seattle. Weights were suspended from the rudder cables where they passed under the spot where the weak floor had been on Flight 427. The weights were increased incrementally from 50 pounds to 250 pounds, but the most this could do was deflect the rudder 3.2°, little more than a full yaw damper deflection.

There were other exotic hypotheses. Suppose a rudder cable had snapped. What would happen? With the same retired airplane, the group tried to simulate the effect of a rudder cable breaking under tension. Acoustics expert Jim Cash was invited to record the proceedings. Using a long-reach, heavy-duty bolt cutter, a powerful tool capable of snapping thick steel cable like a scissors cutting twine, the investigators set to work cutting the cable, then repairing it to repeat the test. A loud bang was heard each time the cable snapped, but beyond that nothing happened. The rudder remained motionless, as did the pedals. When Cash replayed the cockpit voice recorder, he thought that the sound of the cable snapping might resemble one of the mystery sounds on the Flight 427 tape. But the cable-snapping sound had a ringing due to reverberation of the rudder system components; ringing was absent from the Flight 427 recording. A computerized analysis of the cable-snapping test placed the sound all over the spectrum, whereas the mystery sound on the Flight 427 tape had a singular low-frequency quality.

Another what-if involved the possibility that something, perhaps a piece of metal dislodged from elsewhere in the tail fin cavity became jammed in the linkage of the rudder system, just where the input arm enters the PCU servo valve. A folded business card was inserted into the linkage, and the rudder pedals and yaw damper were operated. This produced surprising results. Any operation of the rudder pedals or the yaw damper pushed the rudder to its full deflection. The rudder returned to neutral when the yaw damper or pedal command was canceled, and the business card often fell out. But pushing the opposite pedal as a means of attempting to recenter the rudder only served to keep the business card in place. This test intrigued the investigators until they realized that a large amount of dexterity was required to put the business card in place behind and underneath various arms and levers on the outside of the PCU. On reflection, they decided that, acting under the force of gravity alone, no fragment of sufficient size could reach that spot unless the airplane was turned on its side and shaken vigorously.

The business card exercise highlighted once again the unpredictable results that investigators obtained from tests with the main rudder PCU. In another series of tests on the retired 737, investigators simulated a series of jams on the standby rudder actuator, the part that moves the rudder when commanded by the standby rudder PCU. It is brought into action in the event of a failure of both of the main hydraulic systems. This actuator swings about uselessly when the main hydraulic and rudder systems are in operation but was the subject of a safety recommendation in the wake of the crash of Flight 585. Investigators had discovered that a shaft was rubbing against a bearing surface, transferring metal in a friction process known as galling, and feared that this rubbing might jam the shaft and somehow affect the overall rudder system. Similar galling was found on the Flight 427 standby rudder actuator.

Normally, the yaw damper should have no impact on rudder operation. For example, if there is a 3° yaw damper command on top of a pilot command of 6°, the rudder should not move any more than 6°. In other words, separate rudder commands should not be combined. But when the movement of the standby actuator was artificially jammed or restricted in testing, the rudder commands were often combined. In addition, the rudder often traveled farther than the normal 3°-range of the yaw damper, or else it traveled farther to the left than to the right, or vice versa. Attempts to center the rudder from the cockpit required much stronger pedal forces than normal. The tests left investigators completely baffled.

So much for what happens to the rudder system when you test inside the airplane. What happens if there are sudden or unusual forces acting on the exterior? Imagine a rudder pedal or a yaw damper command moving the rudder, and there's a sudden sideways gust of wind. Or imagine the effect of wake vortices tugging at the rudder. In September 1996, the Systems Group gathered yet again at Boeing in Renton. Bench testing was set up to simulate left and right loads on the rudder, while a range of simulated yaw damper and pedal commands directed the rudder in either direction. Two PCUs were employed, a new unit straight off the production line and the unit from Flight 427.

The group was particularly interested in what might have happened if the tail fin of Flight 427 entered the wake vortices of the Delta 727 flying ahead of it. As the rudder panel was pushed or pulled, what forces were transmitted through the rudder linkages, and how could they affect the PCUs behavior? Tom Jacky's Performance Group had earlier estimated that those forces could be imposing loads of as much as plus or minus 600 pounds on the output shaft that takes the hydraulic power from the PCU and pushes or pulls on the side of the rudder hinge. Assume that a wake vortex force of 600 pounds is coming from the right and the rudder is being pushed right, against the force, by the pilot or the yaw damper. This means that an additional 600 pounds is added to the force the PCU needs to overcome. If, on the other hand, the rudder is being commanded left, the vortex force is aiding it and is adding to the power being transmitted to the rudder motion. The Systems Group tested the PCUs at both 600 and 1,200 pounds, just to be sure. And in one test, they ramped the load up to 3,500 pounds, to be doubly sure. The loads had no impact on either PCU, which easily shrugged off this interference and operated perfectly.

apter 17 BOEING ON THE HOOK

On October 18, 1996, Chairman Jim Hall convened another public NTSB board meeting to approve 14 new Safety Recommendations arising out of Flight 585, Flight 427, Flight 517, and the growing list of 737 rudder incidents. Like most public NTSB meetings, it was well choreographed, the business at hand having been decided in advance. Some of the recommendations had been first drafted as early as February 1995, when Greg Phillips, with Tom Hauteur's backing, was pushing to have the 737 rudder redesigned. But Hauteur's boss at the time, Bud Leynor, the deputy director of Aviation Safety, refused to act on the recommendations because the NTSB had no proof the rudder was faulty. Hall refused to countermand Leynor's decision when Hauteur approached him on the issue.

But Leynor was now retired, and Bernie Loeb had taken charge of the Flight 427 investigation. Loeb had become more aggressive in his stance on the rudder matter because so much had happened in the past year and a half to validate Phillips's original hunches. The 14 items, which Hall took to the NTSB's board and secured approval for, were safety recommendations, pure and simple. Though they made no allusions to the cause of any accident or incident, they left little doubt in anybody's mind as to what the NTSB was thinking: there was definitely something wrong with the 737, and they wanted it fixed. Among the recommendations were the following:

- Boeing should begin "developing immediate operational measures and long-term design changes for the 737 series airplane to preclude the potential for loss of control from an inadvertent rudder hardover."

- Boeing should make plans for a cockpit indicator that would show a 737 pilot which way (left or right) the rudder was positioned, and to what degree. (The only other way the pilot could know was by the position of the rudder pedals.)

- Yaw dampers should be completely redesigned and replaced on all 737 aircraft. Moreover, Boeing should openly state that pilots should switch off the yaw damper immediately upon any uncommanded yawing movement, with this to be an immediate action, not just a checklist item.

- Several recommendations were aimed at the main rudder power control unit. The NTSB wanted proper inspection intervals and a service-life limit to be set for the main rudder power control unit (PCU). According to the recommendations, Boeing should devise a method to alert mechanics and pilots whenever a secondary or primary slide jammed in the PCU. (This was impossible to detect on an otherwise functioning unit unless the jam caused a problem.) Boeing should also ensure that the slides were capable of shearing through any metal particles that got into the PCU and threatened to cause a jam. The FAA should check on all other passenger airplanes to see if they carried similar valves that might be capable of misbehaving and causing flight control malfunctions.

- Two safety recommendations called on Boeing to devise techniques pilots could use to recover from rudder hardovers.

- The FAA, in certifying new aircraft, should cover the possibility that a flight control surface might jam at its full deflection, something the existing rules seemed to ignore.

- Finally, there was a recommendation that all airliner designs, of all makes, be re-evaluated retrospectively in line with the new ruling.

Boeing executives had been preparing themselves for the Safety Recommendations that Hall signed on October 18, and a well-oiled lobbying machine was poised to spring into action. Once Hall issued the recommendations, it was of course too late to lobby at the NTSB, but the FAA, to which the recommendations were addressed, was fertile ground. Before it can act on any Safety Recommendation, the FAA is required to consult with all parties that might be affected by a new rule. And that included Boeing and the airlines.

There was little Hauteur and Loeb could do but await the FAA's response. If the FAA rejected the recommendations outright, then it would reply to Hall by letter. If it decided a recommendation was worthwhile, it would initiate the first step in the process and publish a Notice of Proposed Rule Making in the *Federal Register.* The publication of this notice alerts the public and industry to the draft of the proposed regulation and offers any interested party the opportunity to comment. A generous deadline is set for receiving objections or amendments. If airlines and plane makers object to a proposal and are unable to secure an outright rejection, they will often try to have the proposal watered down or its implementation delayed, citing cost, disruption caused by taking airplanes out of service, or a shortage of parts. If a new rule, called an Airworthiness Directive (AD), is indeed the result, the FAA will often allow another generous period of time for its implementation. Thus, it can take several years for an NTSB Safety Recommendation to find its way to an airplane in service. The rare exception is when the FAA decides that a situation is so hazardous that immediate action is needed and issues an emergency Airworthiness Directive.

Loeb and Hauteur were anticipating a delay of months, if not years, for any concrete reaction to the proposals they had drafted. Nobody could have predicted they would hear from Boeing within days.

Eleven days after Jim Hall signed the 14 Safety Recommendations, a worried group of Boeing engineers and test pilots clustered around a rain-splattered 737, watching its rudder swing back and forth. In the cockpit was a Boeing test pilot. An aeronautical engineer named Ed Kikta was outside in the rain on a gantry erected beside the tail, holding a clipboard listing a series of rudder movements that the pilot was to execute from the cockpit. Through a maintenance panel in the tail, engineers had rigged up a device to restrict the movement of the PCU, the very part that Hall's recommendations had slated for urgent attention. But this nervous gathering had nothing to do with the NTSB. It was prompted by something much closer to home.

In the days following the second set of cold-soak tests at Renton (they started on October 7, 1996), Kikta had been one of the engineers poring over the results in order to fathom what had caused the PCU to jam. It was a carbon copy of the jam experienced at Canyon Engineering Products' Valencia plant. This time, though, unlike the situation in the Canyon tests, where some of the data was lost due to error, Kikta had much more to work from and was more confident of its accuracy. In particular, there were accurate measurements of the operation of the servo valve as hydraulic fluid was forced through the passageways that led into the rudder actuator, the unit that pulled or pushed on the rudder to move it left or right, quickly or slowly. Kikta was cross-checking the results when he noticed something peculiar. When the valve jammed, it was because the secondary, or outside, of the two concentric slides somehow became stuck in the body of the valve. That should not have been a major problem because the valve was designed so that the primary, or inside, slide would carry on to interpret the pilot's pedal commands by directing pressurized hydraulic fluid down the correct set of holes to cause the piston in the actuator to move the rudder left or right. At first glance, the data told Kikta that was indeed happening: despite the jam, a hydraulic command was heading for the rudder. It took a few minutes for the full impact of what he was looking at to hit him. The fluid was going in the wrong direction. On a real aircraft, the rudder movement would be the reverse of the pilot's command.

Kikta's discovery sent shock waves through Boeing's engineering division. His boss, Jean McGrew, the bookish chief project engineer for the 737, ordered a complete review of the test data, and when the review confirmed Kikta's discovery, McGrew called for another series of tests to establish how the malfunction could happen. The tests on the 737 sitting in the rain were being conducted to discover if the reversal could happen outside of the laboratory. The leading theory was that a series of small deviations from specifications (each still within tolerance) in the various linkages leading to the servo valve were adding up and allowing the primary slide to be pushed farther than it should go. The result was pressurized hydraulic fluid being directed through the wrong holes, and the tests on the 737 confirmed this. After the PCU was restrained by a device to simulate a jam, the pilot stomped hard on the right rudder pedal. Kikta watched, horrified, from the gantry as the rudder swung left, the wrong way.

On the previous day—Monday, October 28—McGrew had contacted the FAA's Renton office to warn them that something was brewing. He gave sketchy details at first, told them about the upcoming tests, and said he would report in more detail when the results were at hand. Now he felt that he had no more choices. The day after the tests on the airplane, he led a delegation to the FAA and explained what had happened. His worry, he explained, was that the 737 might no longer meet its original certification requirements. He offered two proposals. First, Boeing would immediately devise a series of checks for airlines to routinely perform on their airplanes to detect any slackness in the fittings that might allow the PCU's primary slide to overtravel in the event of a jam of the secondary slide. Second, Boeing would redesign any faulty parts identified on the PCU and meet the cost of supplying them to all airplanes in service.

On Sunday, October 27, the *Seattle Times* published the first installment of a long, 17,000-word story by one of its aerospace reporters, Byron Acohido, who had closely followed the Flight 427 investigation. From the very start, Acohido focused on the rudder PCU as the lead suspect in the crash. Acohido was very well informed and, despite some errors, had obviously done extensive research. For example, he knew that the PCU had jammed in the latest cold-soak tests at Renton just a few weeks earlier, despite the lack of an official announcement by any of the parties to the investigation. As the daily appearance of each episode of his five-part series brought further misery to Boeing's management, they took cold comfort in the fact that Acohido had yet to get wind of Kikta's discoveries, which remained a closely guarded secret at Renton.

Acohido was a member of a small team of dedicated aerospace reporters at the *Seattle Times* whose work had produced so many aviation scoops as to generate considerable envy among their rivals elsewhere. (Acohido would later win a Pulitzer Prize, among other awards, for his work on the Flight 427 story.) It had not always been like this at the *Times*, the main paper in what was a Boeing town. About ten years earlier, the paper had been ridiculed in the prestigious *Columbia Journalism Review* as a slavish mouthpiece of Boeing, hardly daring to alter a syllable in Boeing press releases. Some senior Boeing executives traced the advent of the *Times*' excellence in aerospace journalism to that stinging rebuke.

Acohido's aggressive, probing style made him a thorn in Boeing's side. The company accused him of bias and refused to grant him interviews or to answer his queries by the time of his October 1996 series. "Acohido practiced what we called the thesis style of journalism," recalled a now-retired senior Boeing figure. "We used to think he would write the story first, then go out looking for facts to suit his thesis." Later, when word reached them that he had been short-listed for a Pulitzer Prize, there was a serious debate within the company about how they might use errors in his articles in an attempt to discredit him. No doubt they had in mind the case of a previous Pulitzer winner who had to surrender her award and was fired when it emerged that the ghetto interviews on which her winning entry was based had been fabricated. But Boeing gave up any plan to discredit a very popular reporter in his moment of glory because it might backfire.

Acohido had other critics. NTSB staffers complained he sometimes exaggerated or misrepresented the importance of some events, and even some members of the Air Line Pilots Association, who were often Boeing's strongest adversaries in the investigation, wished he displayed more balance. "His timing was good but it was coincidental," said John Cox, an ALPA delegate in the Systems Group. For Acohido, Cox added, Boeing was nothing less than "the Great Satan. He sometimes made allegations that were not true, he isolated himself from the mainstream."

Nonetheless, Acohido was the first to bring the complicated and poorly understood facts of the Flight 427 investigation to a wider public, popularizing a story that Boeing earnestly wished would go away. He was generous with his knowledge and even gave helpful background briefings to less talented colleagues from competing media. He also reversed his paper's earlier poor standing with the *Columbia Journalism Review*, winning a bouquet from its editor in late 1994 in the wake of the Flight 427 crash when he exposed the fact that the FAA had caved in to pressure and given airlines five years to install replacement main rudder PCUs in the wake of the Mack Moore incident.

Midway through the appearance of Acohido's series of articles, Boeing's public affairs director wrote a two-page attack on Acohido's journalistic methods and demanded that it be printed on the front page of the *Seattle Times* to coincide with the publication of Acohido's final installment. The newspaper refused to give the attack such prominence, saying it was a criticism of the singer, not the song, but it agreed to publish the letter elsewhere in the paper at a later date. Two days afterward, on Friday, November 1, Charlie Higgins, Boeing vice-president in charge of safety and performance, joined the fray. Boeing had told the *Times* that Higgins would deal with the content of Acohido's articles, but the statement he faxed all but ignored the article and instead sang the praises of the 737's impressive safety record. Again, the paper's editor decided that he would publish the letter, but it did not address any of the technical and safety issues raised by Acohido and therefore did not merit front-page exposure.

Only hours after the letter appeared, Boeing faxed a press release, quoting Charlie Higgins again, stating that Boeing was issuing a service bulletin advising airlines to check their main rudder PCUs for possible faults. After a week of clumsy denials, the company now admitted it had a problem with the rudder, although it continued to insist that its action was an isolated, precautionary measure and bore no relationship whatsoever to the Flight 427 and Flight 585 investigations. This wasn't, of course, how NTSB chairman Jim Hall saw it. That afternoon he issued a statement saying that the latest discovery at Boeing arose directly from tests associated with the Flight 427 investigation.

"While the NTSB supports Boeing's efforts, we are doing a full analysis of this data to find out how it relates to the Pittsburgh and Colorado Springs accidents," said Hall. "I am encouraged that this brings us closer to finding the answer to these twin tragedies." The same day, the FAA announced separately that it was making the safety check proposed by Boeing a mandatory procedure, the subject of a rare emergency Airworthiness Directive.

Until late on Friday, November 1, Jean McGrew never told the NTSB what was happening with the extended tests on the PCU. Nobody at the NTSB knew about Kikta's discovery of the rudder reversal in his study of the cold-soak test data, the subsequent tests on a real airplane, and McGrew's delegation visiting the FAA's Renton office with proposals for action. McGrew faxed the details to the NTSB in Washington. The fax was handed to Ron Schleede, one of the few senior people left in the L'Enfant Plaza building after 5:30 P.M. Tom Hauteur and Greg Phillips were in Pittsburgh.

Hauteur didn't learn of this development until he got back to Washington and received a phone call from Schleede over the weekend and read about it in the papers. He went ballistic. Bernie Loeb was even more furious. Mistrust had been brewing for the past year. NTSB staffers suspected Boeing of withholding information from them, information that was emerging from lawyers for victims' relatives through the legal discovery process. At times Boeing had submitted answers to requests for information but claimed that the information was "proprietary," meaning investigators could not disseminate it or place it on the public docket, effectively restricting its use.

If anybody knew of the NTSB's dislike of being kept in the dark, it was Jean McGrew. The previous November, at the second public hearing on Flight 427 in Springfield, Virginia, Jim Hall had publicly bawled him out over the failure to inform the NTSB about a private Boeing investigation into uncommanded roll on 737 aircraft. He had been chastised earlier, in January 1995, over Boeing's failure to hand over material relating to several Air France rudder incidents. Boeing's John Purvis had learned a similar lesson. Once, when he arrived in Washington for a meeting, he was taken aside by an NTSB staffer and warned that Bernie Loeb had a signed subpoena in his pocket and was going to produce it if Boeing failed to agree to supply a list of items being requested. Forewarned, Purvis made sure to supply everything he was asked for.

As director of air safety investigation at Boeing, Purvis knew that his job included keeping on the good side of the NTSB and all the other air accident agencies he dealt with around the world so that they would continue to invite Boeing to participate in crash investigations. Purvis knew that McGrew's decision to go first to the FAA would cause problems with the NTSB, but he also knew there was no other way. Given a choice between which agency was the more important to Boeing, the FAA won hands down. The NTSB could only recommend something be done. The FAA could ground an airplane—a whole fleet of airplanes! Like any other plane designers and manufacturers, Boeing officials had good relationships with their opposite numbers at the FAA, which enabled the company to make its case effectively during the rule-making process. The grief likely to emanate from the NTSB was tolerable as long as relationships with the FAA was preserved.

Besides, McGrew felt he had discovered a glitch that went to the heart of what made the 737, in the eyes of regulators, safe to fly. The FAA had certified the airplane on the basis of information that Boeing had supplied. Now, some of that information had proved incorrect. Moreover, after Hall had signed the 14 Safety Recommendations in October 1996, Boeing people began making soothing noises to key people at the FAA, suggesting that the recommendations were overkill since there were no significant problems with the rudder. The latest discoveries showed that was no longer true. Boeing had some serious fence-mending to do.

Meanwhile, Vice President Al Gore was looking for some good news. When President Bill Clinton made Gore head of the White House Commission on Aviation Safety and Security earlier in 1996, the goal was to rebuild confidence in US aviation, which had been having a tough time. Not only was there a major crash left unsolved since 1994, but two additional accidents had occurred with horrific loss of life. In May 1996, a burning ValuJet DC-9 buried itself in an Everglades swamp, killing 110 people. And in July 1996, TWA Flight 800, a Paris-bound Boeing 747, exploded in midair with the loss of all 230 people aboard. Other recent accidents added to the loss of confidence. The previous December, an American Airlines Boeing 757 hit a mountain in Colombia, killing 160 people. In the same month as TWA 800, an engine exploded aboard a Delta MD-88 about to take off in Florida, killing 2 people and injuring 5. In February 1996, 87 people on board had a lucky escape at Houston when their Continental DC-9 made a wheels-up belly landing and slid for 6,850 feet before ending up in the grass.

The highly publicized investigation of TWA 800 at times descended into farce as the FBI, rather than the NTSB, took the lead following widespread speculation that a bomb or an antiaircraft missile had downed the airplane. After examining radar records in the hours following the crash, the FAA told the White House that the missile theory was a strong possibility. Later it emerged that what was taken to be the radar track of a missile was just normal interference, or "clutter." In early November 1996, while the public was still digesting the latest announcements from Seattle about the emerging problems with the 737 rudder, a former White House press secretary from the Kennedy administration, Pierre Sallinger, claimed he had evidence that a US Navy missile had hit TWA 800. Sallinger's "evidence" was later revealed to have been lifted from an Internet conspiracy theory newsgroup.

On January 15, 1997, Gore was scheduled to give a closing speech to an international conference on aviation safety sponsored by his commission, and he wanted to be able to impart some good news that suggested his outfit was achieving something. He badly needed a lift. Gore aides contacted the FAA and were told that the FAA was about to start the rule-making process for directives on the 737 rudder. The aides jumped at the chance to include it in the speech. For the first time, after more than a year of dreadful news, the man charged with improving US aviation safety had the opportunity to suggest that something was at last about to happen. White House spin-doctors plugged his speech well in advance and all the networks carried excerpts.

Gore announced that Boeing was to design a new main PCU that would be incapable of reversal, and retrofit it to all existing aircraft over the next three years at its own expense. There was to be a new, more modern and more reliable electronically controlled yaw damper, provided again at Boeing's expense. Weaknesses in the rudder linkages were being eliminated to prevent over-travel, and a new hydraulic limiter was being installed to prevent the rudder swinging too far when the aircraft was at altitudes greater than 1,000 feet. Although two of the four initiatives Gore announced had been recommended by the NTSB the previous October, the NTSB was not mentioned. Instead, Boeing and the FAA got the credit. Gore praised the 737 and the changes that were making a safe aircraft even safer. Boeing was cheered for embarking on this program at an estimated cost of $150 million without waiting for a government mandate, and the FAA was praised for enhancing safety in partnership with industry. "And," added Gore, "they help set a tone for an expanded and more innovative approach to improving safety." In effect, the White House had sidelined the NTSB.

Jim Hall waited a month, until February 1997, before moving to re-establish his turf with a ten-page letter to the FAA. The 737 still wasn't safe and the fixes announced by Gore were not happening fast enough, he warned. In the meantime, there was a possibility that the airplane no longer met its certification requirements. And other safety problems of the 737 needed addressing.

To the embarrassment of the White House, Hall's letter received widespread coverage. Drafted by Hauteur at Loeb's urging, the letter documented what investigators had discovered about the behavior of the PCU when the servo valve jammed. Hall warned that the tests ordered by the FAA in their emergency Airworthiness Directive of November 1 were no guarantee that a valve might not jam some time afterward. He also pointed out that a jam of the secondary slide only became evident during high rate rudder movements.

When the 737 was originally certified by the FAA, the airplane needed to demonstrate that, in the event of a jam of any control surface, it could still be safely landed using other controls. Boeing had assured the FAA in 1967 that control could still be maintained through the use of ailerons in the event of a rudder jam. But since the discovery of what is called the "crossover speed anomaly," Hall's letter pointed out, that was not always true: at 190 knots and in a Flaps 1 configuration, a rudder hardover could not be overpowered by the ailerons. A series of simulator studies, part of the Atlantic City flight tests, and subsequent flight tests by Boeing at Renton had been required to correct an alarming deficiency in understanding how the 737 flew at slower speeds. There was always an assumption that the ailerons were more than capable of overcoming full rudder if the 737 was otherwise being flown correctly. Now it was found that at an airspeed of about 190 knots, a 737 with flaps extended to the Flaps 1 position (the minimum extension for speeds near 200 knots) might not have that margin of safety. The rudder was now known to be capable of overcoming ailerons in that configuration, particularly when it went to the blowdown limit, the maximum deflection of the rudder at a particular speed. By its nature, the blowdown limit changes with airspeed; the faster the plane goes, the stronger the force of the airflow and the less the rudder can swing out. Now, the 737 was discovered to be vulnerable to a rudder malfunction at the crossover point for Flaps 1, the point at which the pilot needs to extend some flap to prevent a stall as the airplane's speed drops.

Flying at or below 190 knots on Flaps 1, a pilot had some chance of recovery if he turned the wheel immediately after an uncommanded rudder hardover and wrenched it to its maximum to make the ailerons oppose the effect of the rudder. But any delay on the pilot's part or a failure to put in full wheel could be the beginning of a disaster. At best, the airplane might hang in a semi-permanent banked sideslip. At worst, it would continue to roll and turn upside down, where the chances of recovery grow increasingly slim. Much depended on the rate at which the rudder went over: slow rates of rudder movement offered the best chance of recovery, but a rapidly swinging rudder offered an extremely narrow window of hope.

USAir was the first to start clanging the alarm bells, and as early as October 13, 1995, wrote to Tom Hauteur, imploring him to examine the consequences of this discovery. USAir said they would give any changes their highest priority. After Boeing ran some flights to validate the assumptions it had made in programming its simulator, it confirmed that the rudder did indeed have more control than the ailerons at 190 knots or below when Flaps 1 was selected. But Boeing denied this was a major issue because the probability of encountering a jammed rudder hardover at that point was so slim as to not be worth bothering about. The Air Line Pilots Association took up the cry the following March, asking Hauteur to have the NTSB determine the angle of attack and the G-load on the crossover point for the Flaps 1 configuration, using figures they were convinced were readily available from Boeing.

But advising airlines how to fly their airplanes was not Hauteur's job, and the problem was lobbed back into the FAA's lap, where it properly belonged. FAA experts scheduled a closed-door huddle with Boeing, ALPA, and the NTSB for January 1996, but no early solutions were forecast. In any event, a frustrated USAir took matters into its own hands in early December 1995. It informed its pilots that they had to fly approaches to landing at least 10 knots faster than usual to counteract the newly discovered crossover speed anomaly. At a higher speed, the ailerons would be able to counteract the rudder should a jam occur.

When USAir officially unveiled its new procedure at an FAA meeting in January 1996, it received the blessing of both Boeing and the FAA. However, despite heavy lobbying from ALPA representatives, neither Boeing nor the FAA would make the new landing speed mandatory for other airlines. The FAA instead advised pilots who experienced a sudden rudder problem to systematically shut down all their hydraulics systems until the problem went away. Boeing's stance was somewhat underscored by the NTSB's reluctance to become embroiled in the issue. But it was not long before the practice started to spread throughout the 737 fleet in the US, helped in no small measure by ALPA's preaching the new gospel.

Hall made another point in his February 1997 letter to the FAA. Certification of an airplane required that all controls be duplicated with automatic backups unless it could be demonstrated that the potential for their failure was "extremely improbable" or was easily corrected by the pilot without needing exceptional piloting skill or strength. Boeing always maintained that the single-panel rudder system had a built-in backup, thanks to the dual valve design of the servo valve. Besides, its failure was "extremely improbable." But in light of recent discoveries, Hall insisted, this was no longer true. NTSB research had uncovered five instances of valves jamming completely prior to 1990 alone, due to wear, corrosion, or foreign matter contamination. Two instances had occurred in flight. Now they knew that a jammed secondary slide in the servo valve could actually cause the rudder to reverse. And it would take exceptional skill on the part of the pilot to fathom that malfunction and recover the airplane.

Hall's letter also alluded to the fact that Boeing claimed "grandfather" rights when seeking certification of the 300, 400, and 500 series 737s in the late 1980s. Although stricter rules were in force in then, Boeing was allowed to rate the new models according to less strict guidelines that were up to 20 years old. Yet, according to what was now known, the airplanes might not meet even those earlier requirements. "The results of the recent main rudder PCU tests indicate that a jamming of a servo valve secondary slide (a single failure) and subsequent reverse rudder operation during a normal pilot response can no longer be considered an extremely improbable or an extremely remote event, and thus raise serious questions about the validity of the certification of the existing B-737 main rudder PCU," Hall wrote.

Hall wanted the replacement PCU program to be accelerated: "The recent tests indicate that the current B-737 rudder system does not provide the same level of safety as on similar transport category airplanes and that the potential of a rudder reversal may be precluded by the installation of the redesigned servo control valve." In addition, he wanted the FAA to warn pilots of the dangers of applying vigorous rudder movements that could cause a reversal if the PCU was jammed, and to ensure they were properly trained to recognize and recover from rudder hardovers and reversals: "A reverse rudder response represents a seemingly implausible event that pilots have no reason to expect and, in fact, is counter to all of their training and experience. Pilots would typically continue to apply pressure on the rudder pedal in an effort to control the airplane. This reaction would only exacerbate the problem and possibly lead to a loss of control."

The FAA's Tom McSweeney, director of aircraft certification, vigorously defended the FAA's actions by rejecting Hall's proposals on the very day they were released to the media. This precipitous public criticism was very unusual behavior, even for an agency that had more than once been accused of cozying up to the plane makers. Normally, a respectable period of grace would elapse before the FAA would declare its opposition or acceptance of such proposals, unless they were to be the subject of an emergency directive. "We believe, as much as we have studied this aircraft and this rudder system, that the actions we have taken assure a level of safety that is commensurate with any aircraft," he said in a prepared statement. And Boeing, in answer to press queries the same day, said it was working flat out to design the new PCU and highlighted the fact that the checks mandated the previous November had still not turned up any jams.

Boeing and the FAA several times repeated the mantra that they were "working to make a safe airplane safer."

Chapter 18 REVERSALS

By the fall of 1996, a few months before Jim Hall made the case against the Boeing 737 in his long letter to the FAA, Bernie Loeb and Tom Hauteur and their colleagues at the NTSB had a fairly good idea of what had happened to cause the crash of Flight 427:

1. Flight 427 encountered the wake vortices of Delta Flight 1083 and rolled.
2. The pilot (or the yaw damper) applied right rudder to correct the roll.
3. The main rudder power control unit (PCU) servo valve jammed.
4. That jam led to a rudder reversal.
5. That reversal was a hardover.
6. It happened at 190 knots, close to the crossover speed for Flaps 1, where the ailerons cannot overcome the effect of the rudder.
7. The aircraft yawed, rolled, stalled, went into an uncontrollable spin, and crashed.

Boeing did not agree with the NTSB's scenario, and insisted on other factors being taken into consideration:

1. There was no evidence the rudder had jammed or moved on its own.
2. Nothing like this had ever happened before in the long, safe history of the 737.

3. Kinematic studies showed a variable rudder movement, not a steady movement toward a hardover.

4. The pilots, startled by the sudden movement caused by entering the wake vortices from Flight 1083 in front of them, might have pressed the wrong rudder pedal and kept it pressed to the floor.

5. There is evidence from the literature that startled people make mistakes.

In a remarkable 29-page document conceived by Jean McGrew, Boeing's chief project engineer for the 737, and titled "Boeing Contribution to the USAir Flight 427 Accident Investigation Board," the plane maker swept aside any other evidence as irrelevant and concluded decisively that the airplane crashed because the pilot put in full left rudder and held it there after encountering the wake of Flight 1083.

The Boeing submission outraged NTSB investigators because it ignored the emerging evidence that pointed to a problem with the main rudder PCU, including the cold-soak tests, and it ignored protocol, which called for submissions to be made directly to the investigative staff. Instead, Boeing dispatched copies over their heads to Jim Hall and the other four NTSB board members before their October 18 vote on the 14 new Safety Recommendations, mostly related to potential rudder problems. Boeing obviously wanted to nip this action in the bud.

There is little doubt that Boeing personnel believed that the investigators were determined to find a mechanical fault with the airplane, contrary to all the evidence the plane maker believed it had produced. Privately, they accused the NTSB of being in some sort of unholy alliance with ALPA, the all-powerful pilots' union. Senior Boeing personnel who worked on the investigation genuinely believed (and some still do) that pilot error was to blame, but they claimed the NTSB was prepared, for undetermined and possibly political reasons, to allow ALPA and USAir (which also wanted a mechanical scenario) to dictate the course of the investigation. Yet when preparing the document, Boeing executives abandoned any pretense of a balanced review of the evidence and did themselves few favors in the process. In the introduction to the document, Boeing stated:

> ... While there is no service history or NTSB Systems Group finding to support a belief that a full uncommanded rudder input occurred during Flight 427, there are data and reported accidents, incidents and events that explain commanded rudder inputs during the Flight 427 upset....

Later it added:

> ... We know for a fact that the crew made full column inputs. The analysis has shown that full wheel inputs were made. The analysis had shown that two large rudder inputs were made as well....

A casual reader might believe the investigation had indeed proved that the crew made those rudder inputs. But no such proof existed. In its reconstruction of the events leading to the accident, Boeing said unequivocally that the pilots' putting in full left rudder and keeping it in caused the accident. No evidence, other than a Boeing claim that this had occurred elsewhere, was presented for this supposition.

In addition, Boeing claimed that the ailerons were overcoming the full left rudder, but that the pilots precipitated a stall by pulling back too soon on the control column before the wings were leveled. In fact, following flight tests, Boeing personnel were fully aware that at a speed of 190 knots in the Flaps 1 configuration, the rudder probably had more authority than the ailerons because of the crossover speed anomaly. The pilots' actions may indeed have precipitated a stall, but there was no reason to believe the ailerons were overcoming the rudder—the plane was rolling, and there was apparently nothing the pilots could do about it.

Reviewing the Flight 585 accident, the Boeing document included a quote former NTSB chairman Carl Vogt made at the board meeting that signed off on the final report on Flight 585: "I find a lot of evidence that there was a rotor there that they got into that flipped them over." Then Boeing added that the NTSB "could not identify conclusive evidence to explain the crash." Boeing's selections from the 585 report mentioned only rotor evidence, completely ignoring the problems found with the PCU and other rudder parts during that investigation. In fact, in the conclusion of the Flight 585 report, the NTSB clearly stated: "Either meteorological phenomena or an undetected mechanical malfunction or a combination of both could have led to the loss of control."

Boeing's sanitized version of the evidence enraged both USAir and ALPA. In a letter to Jim Hall, USAir called it "highly inappropriate, inconsistent with NTSB regulations and Guidelines, and contrary to the Board's direction to the parties." The content was "little more than wild speculation based on dubious conclusions drawn from flawed and unsubstantiated data." That letter from Malcolm Armstrong, USAir's Vice President of Corporate Safety and Regulatory Compliance and the airline's new accident supremo, called it "an untimely and improper attempt to influence the probable cause finding in this accident before the investigation is complete." The document, Armstrong added, appeared to be drawn mostly from Boeing's kinematic study. "USAir believes that a careful review of that study will reveal that it is fundamentally flawed and unworthy of serious consideration by the board."

Armstrong took three pages to demolish the Boeing document, while ALPA did it in two short paragraphs:

> ALPA believes that the *Boeing contribution to the USAir Flight 427 Investigation* [sic] is their analysis and therefore a submission. Therefore it should not be used by any investigative group. In reviewing the document, we find that it relies on a flawed draft kinematics study that was produced solely by the Boeing Company without input by the Aircraft Performance Group. We believe that Boeing's attempt to influence the Safety Board with their single-party analysis is premature.
>
> ALPA believes that this submission makes no contribution to the fact finding portion of the investigation.

The NTSB ignored the Boeing submission when acting on the Safety Recommendations that it first discussed on October 1 and passed on October 18, 1996. Although it was a short-term failure for Boeing, in the long term the document helped send a signal to the NTSB that, notwithstanding the progress it had made in detecting problems with the 737 and its components, nothing had emerged to link those problems with the crash of Flight 427. Already, Hauteur and Loeb were getting almost weekly protests from Boeing about aspects of the investigation, in particular the failure of the Human Factors Group to take the plane maker's ideas more seriously. If that was the case now, what would it be like if they came up with a probable cause for the accident without a "smoking gun" to back it up?

Up to this point, the NTSB had bowed to the superior knowledge and expertise of Harry Dellicker, the Boeing kinematics expert who had painstakingly built up a compelling reconstruction of Flight 427's last moments in the air. Although USAir and ALPA were quick to dismiss kinematics, NTSB investigators could not do so without hard evidence to confirm their own theory of the 427 crash. Dellicker's kinematics had enabled him to suggest at precisely what moment the various controls were deployed. He appeared to be able to isolate and correct errors and anomalies in the data taken from 427's flight data recorder.

Dellicker's kinematics compensated for the fact that aircraft performance measurement is far from an exact science and the output of the flight data recorder of a doomed aircraft tumbling through the sky is not always reliable. In such circumstances, even basic readings like altitude, pitch, and roll can be inaccurate. Because an altimeter, which measures altitude, is based on a barometer and measures atmospheric pressure, it may give false readings while the aircraft is in the low-pressure zone at the center of a wake vortex. It may also be affected by high pressure caused by the aircraft being in a dive. The measurement of a pitch more than 90° nose down or a roll more than 80° can also be erroneous because the instruments are meant to indicate what happens in normal service, and may not be capable of such extreme measurements. Data sampled at rates less than two times per second in a fast-moving aircraft are especially unreliable because a lot of unrecorded action can occur when nothing is being measured.

Dellicker had devised ways of identifying what in the data was an accurate measurement and what was a random disturbance, or "noise." With his computer, he was able to pull erratic bits of data out of the margins and put them into the main picture, where they belonged. Or he could decide to ignore something as hopelessly lost data. He would take the chunky graphs of his flight parameters and patiently iron out any irregularities, turning the jerky, flapping, improbable motions of a wing into a smooth, rolling motion, or the furious nodding of a nose into the progressive up and down pitching of an aircraft. He achieved this by interpolation—actually inventing missing data and, by a process of trial and error, putting it where it appears to fit on a curve. That, in turn, enabled him to calculate other missing data.

The exercise was made even more complicated by the possible impact of the wake vortices of Delta Flight 1083 and the poorly understood sideslip of a 737 as it crabs and skids under the influence of its large rudder. Trying to estimate how the wake vortices could affect the complicated shape of a 737 airframe was a gargantuan task, even for the talented Dellicker. The findings of the Atlantic City tests had produced a more accurate picture of a wake vortex, how fast it revolved and how big it was. But no one had yet worked out the forces on a complete airplane. Dellicker came up with an ingenious solution for the problem: he simply chopped up the airplane. On paper, he sliced the plane up into manageable chunks, worked out the vortex forces on each chunk, then added the results together. It provided a remarkably accurate simulation of what actually happened.

Next, he had to decide what combination of flight controls could cause what had happened—which aileron on which wing, how much rudder and in what direction. He knew about just one flight control: the elevators, controlled by the fore-and-aft movement of the pilots' control column, was the only flight control directly measured by the flight data recorder.

Like the NTSB's studies, Dellicker's kinematic exercises confirmed that only the rudder could have produced 427's motion from the point when the airplane entered the wake vortices until it started to stall—a total of about 10 seconds. More than that he could not tell because the behavior of a stalled, semi-inverted airplane was a blank book, even to Boeing.

If Dellicker's analyses were accurate, it meant the NTSB had been building its scenario of how the pilots lost control of Flight 427 on false premises. The theory that Hauteur and Phillips had been developing in late 1996, following the remarkable results of the cold-soak tests on the PCU, assumed an airplane that experienced a jam followed by a reversal of the servo valve, which led to an uncommanded rudder hardover. A rudder hardover of this type is likely to display a linear movement; in other words, once the rudder starts to move, it would continue to move at a steady rate until it reaches its fullest extent. They envisioned no stopping or changes in the rate of movement.

But Dellicker had produced a rudder movement that varied. His study appeared to suggest that it could only be produced by human action. According to this scenario, at 2 minutes and 58 seconds past 7:00 P.M., Flight 427 entered the wake vortex as it was coming out of a left turn. The vortex caused the airplane to yaw and roll left, so the pilot put in considerable right wheel to counter it with the right aileron. This corrected the left roll and yaw, but also produced a right roll and yaw. The pilot then put in 12° of left rudder. One second later, as the rudder and the ailerons were returning to neutral, he put in the left rudder again and held it down all the way until impact. He also held the right aileron to oppose it.

By now, Greg Phillips had collected sufficient information from tests and other incidents to suggest that the rudder system on the 737 was capable of erratic behavior. At the same time, feedback from the pilot community ridiculed the idea that a pilot would not only apply wrong rudder, but hold it in, as Boeing was emphatically suggesting. Was it possible that Boeing was misinterpreting its otherwise impressive data? Could there be another explanation?

Dennis Crider was given the job of finding out. Flight 427 was his first accident experience with the NTSB, but the fledgling investigator had an impressive background in flight simulation and in aircraft stability and control, gained from working at McDonnell Douglas. Crider decided that his first task would be to see if Boeing's approach held water. He set to work devising his own experiments with a view to validating Boeing methods. Dellicker's work impressed him, especially little touches of genius like chopping up the airplane to model the impact of the wake vortices produced by Flight 1083. He was envious of some of the resources at Dellicker's disposal. For example, Dellicker had a so-called "smoother," a computer program that could iron out inconsistencies in the data. "If you sample once per second, you cannot deal with things that happen faster than that. That's all about smoothing the data; it's what we call conditioned data, and at the time Boeing had a better smoother. In fact we didn't have a smoother at all," Crider recalled.

Although he couldn't fault Dellicker's overall method, he did distrust the use of kinematics for short-term events like the Flight 427 crash, where the time frame under study was only 10 seconds. Crider's own preference was an old-fashioned simulator study. Using trial and error, the investigator would attempt to "fly" the simulator through the data points on the flight data recorder printout, looking to see what use of flight control surfaces provides the best match with the recorded data. One reason for his wariness was Dellicker's need to use data estimates to compensate for the low sampling rates of parameters such as heading, which was measured just once per second. When you guessed where the nose of the airplane might be pointing a half-second later, as Dellicker had to do, you could be "spot on"—i. e., right on target—or you could be off by several critical degrees. Either way, Crider thought the kinematic approach meant making educated guesses on rudder movement.

"There were various curve fits you could use in that kinematics extraction to match the motion of the airplane," he explained. The problem was that the data was only recorded once per second, so for a kinematic analysis Dellicker had to do a curve fit to supply data between those points—in other words, filling a blank space in a curve with a segment that would blend nicely into the curves on either side of it. "There are mathematical ways of doing that, but it might not be what an airplane could do," said Crider. "If the one-second gap doesn't matter, as it won't for a long-term issue, then a kinematics extraction is fine. It's still fine for a small thing; but you run into fewer errors with a simulation. You know your airplane is going through the points. Basically the kinematics validation had noise in it. Whereas with simulation, we didn't have that noise."

When he had finally mastered the Boeing kinematic techniques, Crider headed for Renton to pore over Dellicker's work, line by line. Boeing gave him an office and a computer, and he set to work.

One day, working with Dellicker and experimenting with the Boeing software that fitted curves to estimated data points on the graph of airplane motion, he tried a different approach on a double-humped curve near the beginning of the Flight 427 accident data. In the Boeing analysis, the rudder remained deflected to the left but it nonetheless moved five times, sometimes back toward neutral, in the space of 3 seconds. When plotted on a graph, that analysis produced a characteristic double-humped shape formed by two peaks, side by side, one smaller than the other. In the NTSB version of events, the rudder moved straight to its maximum deflection limit without hesitation, achieving the blowdown limit in less than 1 second. At the point where Boeing's graph had two peaks side by side, the NTSB's graph had an almost vertical cliff. The Boeing version suggested variable inputs, more likely to be pilot-induced, while the NTSB version showed linear motion—suggesting a single, steady, uninterrupted input that could be more easily explained by a mechanical malfunction such as a jam of the main PCU.

Crider was curious to know if the mathematics of Boeing's method could come up with the same answer as the NTSB's method. Could a slightly different approach to the problem smooth out that characteristic double hump and turn it into a cliff? He took the figures for the airplane's compass heading and ran them through the Boeing computer program. To Crider's amazement, it produced a result very similar to the NTSB's. Instead of a double hump, there was a cliff.

> ... I was taking the heading and going between the known heading points and moving the curve fit between those with the tool that they had so you can shape between the points ... Then we thought, what would it do if we did this differently? And we found that if you ran the kinematics again you got a fairly linear rudder. The double hump didn't exclude a mechanical thing but something more mechanical would tend to be linear. Harry and I looked at it and we thought, Hey! There's one half of the puzzle. Later, we found that the jam scenario matched it. It was a surprise bordering on shock when we saw it, especially for Boeing....

Dellicker took the results of this new approach to his superiors, who were shocked by the finding and its implications. Boeing could not deny the validity of the new linear interpretation of the kinematic approach, although they would insist that both solutions were valid.

As far as Crider could see, Boeing was prepared to go along with the new direction he had mapped out. He encountered no attempt to persuade him that his work was flawed. Back in Washington, few appeared surprised when Crider announced what had happened. It was as if they suspected as much all along.

Asked if he thought that Boeing had deliberately skewed the kinematic analysis to achieve a desired result, Crider kept an open mind. "There was no conscious effort at that time," he said. "We were working with pretty good people at that level of the organization. We're talking about engineers here. We worked with data and we were looking at a new direction."

USAir was not as tolerant. The following April, its director of flight safety, Captain George Snyder, accused Boeing of manipulating the data to support a predetermined result. In a letter to Aircraft Performance Group Chairman Tom Jacky, Snyder said the reason he had been given by Boeing for manually manipulating the data was to "reduce the sinusoidal noise observed in the derived rudder time history." But that noise, said Snyder, was a direct result of the low flight data recorder sampling rate, and he could not see what purpose was served by removing it. "Moreover, even if reducing this noise was appropriate, the particular manipulations chosen by Boeing to eliminate this noise creates what US Airways believes was a false suggestion that there was a double rudder pump."

Crider had made a significant move in carrying the NTSB's case to Boeing and proving there could be an alternative scenario to the pilot error theory. But he still had a long way to go to conclusively prove his point. Before he could present a more accurate reconstruction of rudder and aileron movements, he needed to know two things. First, what are the forces acting on the rudder from the servo valve pushing it to a hardover? And second, despite a jam, could a pilot reduce the extent to which the rudder was turned during the hardover by pressing hard on the appropriate jammed rudder pedal?

Although Crider was attached to the Aircraft Performance Group, and virtually all of his investigations involved the movements of control surfaces—the mechanics of the airplane—it was almost inevitable that over the coming year he was to work very closely with two other investigators, Human Performance Group Chairman Malcolm Brenner and Systems Group Chairman Greg Phillips.

The Eastwind Flight 517 incident had demonstrated how hard it was for a pilot to make much headway against a rudder pedal during a jam. Captain Bishop failed to center his rudder pedals, despite lifting himself up about 2 inches off his seat, which would have increased his leg force by more than 30 percent, according to an NTSB estimate. But Bishop was only 5 feet 7 inches tall. By contrast, Charles Emmett, the Flight 427 pilot, was a fit man standing 6 feet 3 inches tall and weighing 210 pounds. If the rudder pedal could be budged at all, he was one to do it. (Later, Boeing was to ask if he was not too tall for the cockpit, which had been designed for a person up to 6 feet 2 inches tall.)

However, Boeing highlighted a possibility that both Flight 427 pilots had been pressing hard on the wrong rudder pedal, the left one, thus preventing the plane from recovering control. When the wreckage was examined, both left rudder pedals were found to be fractured, the right ones merely bent. Malcolm Brenner asked the deputy armed forces medical examiner, Dr. David Hause, for his opinion. Hause replied that the fractures indicated both pilots were likely to have been applying strong pressure on the left pedal, possibly by putting most of their body weight on the left foot with the knee in a locked position. More definite proof might have been available from a postmortem examination because he would expect to see what military aviation pathologists call "control fractures." These include mid-foot fractures, collapsing or telescoping of leg bones, and hip fractures. Unfortunately, not enough remains of either pilot were recovered to conduct a meaningful forensic examination. "This makes this scenario a possible explanation," Hause wrote to Brenner, "rather than an opinion with quantifiable probability."

In June 1997, during an all-party meeting in Seattle, Boeing provided demonstrations of rudder jams on a newly manufactured 737-300 that had been outfitted with a special tool to simulate PCU jams at different positions. By this stage, more had been learned about the behavior of the PCU during jams of the secondary slide, in particular how jams at different positions produce varying results.

Boeing test pilot Mike Carriker, taking a break from his Human Factors Group role, coordinated the demonstrations. Brenner sat in the right seat, where the first officer normally sits, with Ben Berman (the new Operations Group chairman) in the left seat. Because of his height (6 feet 3 inches, the same as Flight 427 pilot Emmett), Brenner had to place the seat in the full back position for legroom even though his leg length was less than that of the Emmett.

The first demonstration had the secondary slide jammed at about 25 percent from the neutral position. Brenner pushed each rudder pedal slowly to its full down position as through performing a slow rudder system check. He thought the right pedal seemed easier to push than the left pedal, although the difference was subtle.

But the response was more dramatic when he pushed a pedal hard. In five out of seven instances, this triggered a reversal, and he felt the rudder pedal come back up. "The motion was slightly slower than an input I would have expected from a human being," he noted. "The motion was steady and continued without pause no matter how hard I pushed to counter it. 'Unrelenting' was a description that, at that time, seemed to capture my impression."

After the left pedal reached its upper limit in one instance, Brenner released his pressure on the pedal to stop fighting its upward motion. To his amazement, the reversal of the rudder system ended almost immediately, and the rudder pedals returned to their neutral positions. "On subsequent trials, I stopped fighting the rudder motion earlier before the left pedal had reached the upper stop. Again, the rudder motion stopped almost immediately as soon as I stopped applying pressure, no matter where the pedal was located, and the pedals returned to neutral."

The second demonstration represented a jam of the secondary slide at the neutral position. Brenner felt a slight difference between the two pedals, but any resistance was easily overcome and there was no reversal, no matter how hard or how rapidly he pushed.

The third demonstration represented a jam of the secondary slide at about 50 percent from the neutral position. Any abrupt motion on the pedals initiated an immediate rudder reversal. If he pushed them slowly and steadily, Brenner could usually move the pedals to their stops without starting a reversal, but even this slow pushing sometimes initiated a reversal, with the right pedal moving to the upper stop. He discovered the rudder reversal was faster than with a jam in the 25 percent position. "It was impossible to stop the motion by physically pushing against the rudder pedal. On several trials, I tried relaxing my input momentarily before the rudder pedal reached the upper stop. I found that the rudder reversal motion continued. This had not been true with the 25 percent jam, when the relaxation of pressure seemed to automatically stop the reversal motion. This motion was faster, easier to initiate, and more difficult to stop."

Brenner had discovered that the rudder could behave differently, depending on where the servo valve jam occurred. It could move faster or slower, go left or right. A jam of the secondary slide at the neutral position, where the rudder is aligned with the tail fin, produced no reversal. But jams at other positions produced different forces on the rudder pedals and different speeds of rudder movement as it moved toward a hardover. A jam of the secondary slide at 55 percent from the neutral position would result in the rudder moving at 15° per second. A jam at 100 percent—full travel—would see a rate of movement more than twice that. Brenner also learned that a jam could sometimes be cleared simply by ceasing to fight the rudder. Left to its own devices, a reversal might go away. Or it might not.

Ben Berman had similar experiences and agreed with Brenner's observations. He tried a test of his own during the third demonstration. As the rudder was reversing and the pedal traveling remorselessly upward, he leaned across and flicked the hydraulic system switch to standby rudder. The reversal action ceased immediately, and he could center the pedals easily with his feet. During subsequent rudder movements with the standby rudder system engaged, Berman found the rudder system did not reverse.

Dennis Crider was intensely interested in the rudder jam demonstrations. His reconstruction of Flight 427 in the simulator was far from perfect, and there were still some interruptions to the smooth linear movement of the rudder. Brenner had a theory that answered. On the cockpit voice recorder tape, Emmett is heard to grunt on several occasions, possibly with exertion. What if, suggested Brenner, Emmett was grunting because he was trying with great effort to center his rudder pedals? When Crider looked at the recorder transcript, he realized that some of the interruptions of rudder movement occur just as Emmett is heard to grunt. He found that the timing of the grunts and the interruptions of the linear movement of the rudder were almost perfectly matched.

In another puzzling aspect of his simulations, Crider's rudder appeared to be moving too far, farther than predicted by Boeing's blowdown limit table and the first officer's attempts to take corrective action. When the rudder reached a certain point of travel, the airstream created by the airplane's movement should have prevented it from swinging out any further. Later he discovered why: the Boeing table failed to take sideslip into account. As the airplane yawed, the rudder (but not the tail) tended toward alignment with the airstream, thus reducing the blowdown effect and allowing it to move a degree or two farther out to about 12°.

Meanwhile, Greg Phillips was working out what happened when the yaw damper was activated during a jam of the secondary slide. He set up bench tests using the original PCU from Flight 427 and discovered that a reversal could be induced with some types of jam. Was there anything the PCU was not capable of?

Chapter 19 SUBMISSIONS

In August 1997, Tom Hauteur finally called for submissions from the parties to the investigation of Flight 427. Apart from helping with the fact-finding phase of a crash probe, parties are given the opportunity to offer their analyses of those facts and suggest likely causes (or "probable causes," as the NTSB phrases it). Hauteur's call for submissions was unusual, however, because the fact-finding part of the probe continued. The Systems, Human Factors, and Operations groups were still at work. Not surprisingly, Hauteur's request led many observers to assume that one of the longest-running crash investigations in the NTSB's history was finally winding up, with a finding expected within months, if not weeks. They could not have been more wrong.

In its submission, Boeing wasted little opportunity to reprise its controversial theories. It returned to the pilot error theme that it first introduced a year earlier in the document titled "Boeing Contribution to the USAir Flight 427 Accident Investigation Board." This time, the manufacturer paid more than passing attention to the work of the Systems and Human Factors groups before dismissing them as irrelevant. Drawing support mainly from examples and studies from outside the Flight 427 investigation (there were 100 appendices in the submission), Boeing insisted there was sufficient precedent to suggest that the pilots could have not only stepped on the wrong pedal, but held it down until impact. Among the few items relating directly to Flight 427 research was a selection of facts from the company's own kinematic study. Although Boeing conducted this study independent of the main NTSB investigation, it did cite Dennis Crider's work, although the double hump of variable rudder movements, not the smooth, linear movement favored by Crider, was featured in the graphs it presented.

Dismissing suggestions that the rudder power control unit (PCU) might have misbehaved, Boeing's submission quoted the very first Systems Group report, produced in 1994, as evidence that the unit "is capable of performing its intended function" and "was incapable of uncommanded rudder reversal, or movement." The submission added, "While other 'reversal' failure modes were later identified, nothing in the analysis or testing conducted after these findings were released has provided any physical evidence to the contrary." It was an amazing glossing-over of the huge body of work conducted by the Systems Group into the numerous ways the PCU aboard the 737 could misbehave, although Boeing was correct to say that no smoking gun had been discovered.

Boeing then moved into human factors territory, but paid only scant attention to the real work the Human Factors Group had accomplished and collectively approved. In some cases, the very fact that Boeing had mailed the group an academic report, no matter how unrelated to the matters at hand, was sufficient for the plane maker to claim the report as part of the formal investigation. An example was a study of incidents of unintended acceleration among motorists who had stomped on the gas pedal instead of the brake. In its submission, Boeing cited this as evidence that the pilots could have pushed the wrong pedal. The report had not been commissioned by the Human Factors Group, was angrily rejected by at least one member of the group as irrelevant, and was not included in any of the group's formal reports. Yet Boeing insisted on passing it off as a product of the investigation.

Other studies, including some that examined incidents where pilots had actually pressed the wrong rudder pedal, were far more relevant. But even these studies had not won widespread acceptance by the investigators as having a bearing on Flight 427—with the exception of the 1994 crash of a Sahara Airlines training flight in New Delhi. This case was examined closely by the Human Factors, Aircraft Performance, and Systems groups until late in 1997, when the Systems Group found what appeared to be a makeshift part in the servo valve. It was an end cap that may have been re-machined from an old Boeing 707 part. Given the sensitivity of the 737 servo valve to overtravel, it was a significant finding. Although the Sahara servo valve was not fully tested by the NTSB, the finding was sufficient to make investigators lukewarm on the theory of pilot error even in that crash.

Boeing's explanation for the June 1996 Eastwind Flight 517 incident was that the crew had been startled and overreacted by applying excessive rudder. No reference was made to Captain Bishop's claim that the rudder pedals had moved on their own and had been stiff, or to Jim Hall's statement (in his February 1997 letter to the FAA) that there had been an uncommanded rudder movement at the start of the incident.

Boeing's leading thesis was that crews "sometimes react to startling events by making errors in control manipulation." Paradoxically, in its conclusions Boeing admitted that "under the standards developed by the NTSB, there is insufficient evidence to reach a conclusion as to the probable cause of the rudder deflection." Boeing appeared to be hedging its bets. If it couldn't get a pilot error verdict, then an "undetermined causes" result would do, as had happened in the case of Flight 585.

Boeing had the longest submission to the proceedings, and the FAA had the shortest—only two pages. It said that, without firm evidence, the FAA could not support the theory that there was a mechanical reason for the Flight 427 rudder going to hardover:

> ... While the FAA acknowledges that some failure modes of the main rudder power control unit servo valve have been discovered during this accident investigation, it has not been substantiated that any of these failures occurred on the accident aircraft....

Parker Hannifin's short submission highlighted the fact that despite its age—it was manufactured in 1987—the PCU from Flight 427 had passed numerous tests at their facility: "In sum, after years of one of the most critical examinations in aviation history, there is no evidence that the main rudder PCU from Flight 427 malfunctioned or was other than fully operational."

ALPA did not hesitate to point out that the accident resulted from a PCU jam causing an uncommanded left rudder hardover. There was, its submission stated, no evidence to support the hypothesis that the pilots applied the wrong controls or that they were startled. On the other hand, ALPA continued, the investigation had discovered ways the PCU could produce an uncommanded rudder movement. Furthermore, the timing of the first officer's grunts heard on the cockpit voice recorder tape coincided exactly with the timing derived from the flight data recorder and the performance simulations of his energetically applying right wheel or attempting to center his rudder pedals to correct a hardover.

USAir concentrated the first part of its submission on what it believed did *not* cause the crash. The crew, the airline declared, was not startled by the wake encounter and did not apply the wrong controls. The probable cause was "an uncommanded, full rudder deflection or rudder reversal that placed the aircraft in a flight regime from which recovery was impossible using known recovery procedures." A contributing cause, it added, was Boeing's failure to advise operators about the crossover speed hazard.

After receiving the submissions, Hauteur was impatient to start drafting his own report. But Bernie Loeb urged caution. He feared that a report containing a probable cause that could not be substantiated would be rejected by the NTSB's board or made the subject of a rare formal appeal by Boeing. This was already happening in the case of the American Eagle ATR-72 crash at Roselawn, Indiana, in 1994. The French government agency that certified the French-made turboprop was formally seeking a reconsideration. There had also been objections to the Roselawn report from the FAA, which objected to Loeb's perceived tendency to fault the agency.

Many NTSB staffers, accustomed to a good working relationship with industry and the FAA, noted a deterioration in some relationships once Loeb became involved. Whereas others might have thought it important to cultivate strong relationships with airlines, plane makers, and regulators, Loeb tended to adopt a less personal, more independent approach and was not slow to show irritation if anybody failed to supply the cooperation he felt the NTSB deserved. Loeb himself later described how he felt:

> ... There were times in which the NTSB's top staff people attempted to work very closely with the industry and the FAA, and by that I mean not to ruffle feathers, and to accommodate them, but that certainly was not my style. I was in an extremely responsible position, paid extremely well to do the job, and we owed the public something, we were the public's protection against unsafe things not fixed. Our job was one which, if we were doing it properly, meant we were not going to be looked upon very kindly by the FAA or the industry. So we had some difficult times throughout the period....

Blaming lax regulation was nothing new for the NTSB, but Loeb made it a hallmark of his tenure and appeared to go out of his way to emphasize any failing by the FAA. If, for example, an airline was chastised for sloppy practices, Loeb wanted to see the FAA's failure to properly supervise the airline listed as a contributory factor.

The FAA did not take kindly to being fingered for blame by a sister federal agency. At the same time, there was often deep anger among NTSB investigators when they saw a repeat of a fatal accident following FAA failure to act on NTSB recommendations. The investigators saw this failure to act as yielding to lobbying by airlines or plane makers, with which the FAA had a close relationship. The 1996 ValuJet crash into the Florida Everglades was a case in point. Years before, the NTSB had recommended the installation of fire and smoke alarms in cargo holds, but the FAA refused to make them mandatory after industry objections. The Roselawn ATR-72 turboprop crash due to icing was followed in January 1997 by the crash of a Comair Embraer turboprop, also due to icing, and the death of all 29 aboard near Monroe, Michigan. Loeb personally directed both investigations, which found the aircraft had crashed because they had been certified to fly at too low an airspeed in icing conditions and had stalled when their speed dropped just a few knots below the safety threshold.

Loeb came under intense personal pressure from the FAA after both reports and the Roselawn report practically spawned a diplomatic incident with the French Government (the ATR-72 is a French-built airplane). The intensity of the disagreement was at least partly reflected in Loeb's language, when he later recalled the incident:

> ... In Comair we basically said it was the FAA's fault and I had FAA people say to me it was the most irresponsible thing they ever saw us do. I don't agree. I think it was probably one of the high water marks of the board. There was an opportunity to have known how this airplane handled in icing conditions if recommendations to change the certification of the icing envelope we had made years before had been acted upon. But they hadn't. With Comair it could be argued, with some merit, that the airplane should not have been flown under 160 knots. His airplane went to hell at 155 knots. The point is, our system is so intolerant of errors that if a pilot loses 5 knots people will die. If that system is OK then what you need to do is plaster a notice on the side of the airplane: *"Hey, if you get into this airplane in icing conditions and the pilot happens to not pay attention for a 2 or 3 second period of time and he loses 5 knots (that's 3.125% of his airspeed!) you will all die."* Then I doubt if very many people will climb aboard. My view was that the system had failed that pilot and, more importantly, had failed the public. Yes, the pilot had not done exactly what he should have done, but that airplane should not have gone to hell and people died 5 knots below the speed the FAA insisted was still an appropriate speed. I guess the FAA and the industry weren't really thrilled with some of the things that I had to say....

Jim Hall was also concerned about the NTSB board's reaction to Hauteur's findings on the cause of Flight 427's crash. He wanted a unanimous verdict of the board when the time came. After investing so much of his public capital in this accident over the preceding three years, he didn't relish a result that would show a split on his board. Loeb, on the other hand, would have been happy with a simple majority verdict from the board's five members. Hall's February 1997 letter to the FAA was a sure giveaway that certification issues would be in store in a final report. The pressure on the board intensified as Boeing invited each member to Seattle for a ride in the 737 simulator to let them see how the company's version of events worked.

Loeb regarded Hall's desire for a unanimous, uncontroversial verdict on Flight 427 as playing politics with board decisions, and he suspected Hall's links with Vice President Gore might have something to do with it. Hall had been a Gore aide in Tennessee, and Clinton appointed him NTSB chairman on Gore's recommendation. But Hall's February 1997 letter lambasting the FAA proposals, which had been announced by Gore, had cut the ground from beneath his political patron. Although Hall felt he was independent of political influence, he still seemed to feel an obligation to conclude the investigation with a non-messy verdict.

Relatives of people who died aboard Flight 427 and their lawyers were expecting a verdict from the NTSB on an almost daily basis during the early weeks of 1998. They were to be disappointed. The TWA Flight 800 investigation was making huge demands on the NTSB's resources. Hauteur was continuously revising his rough drafts of the Flight 427 report to take new discoveries on the aircraft performance front into account. And Hauteur and Loeb were running into serious trouble securing agreement from the NTSB's board on a draft.

The investigation was into its fourth year and they still had no smoking gun. Their first draft created serious problems with two board members, George Black and Bob Francis. Both thought the evidence too circumstantial and condemned the draft report for implying that the 737 was unsafe without producing any hard evidence.

Francis went to Hall and said he could not support the Flight 427 report as drafted. Hall, now willing to settle for a simple majority verdict, agreed to allow Francis to negotiate directly with the staff to see if the report could be revised to his liking. This was a major concession to Francis and was to give him considerable leverage over the shape of the final Flight 427 report.

A Clinton appointee to the NTSB board in January 1995, Francis had been a long-term staffer at the FAA and was its representative at the US embassy in Paris when the American Eagle ATR-72 crashed at Roselawn. He went to the Aerospatiale plant in Toulouse with an NTSB delegation for briefings on the ATR, but later excused himself from the investigation. He became familiar to millions of Americans after the TWA 800 disaster because of his daily news conferences as the NTSB's on-scene board member. However, it was a mission plagued with confusion, and Francis was later accused of surrendering too much authority to FBI agents at the scene. The FBI initially excluded NTSB investigators from critical interviews with airline and airport staff, and even eyewitnesses to the disaster. Francis was also accused of making decisions that were best left to Al Dickinson, the veteran investigator-in-charge, and he went to his former colleagues at the FAA for advice, not the NTSB, when it appeared the FBI would fully take over the investigation. However, his handling of on-scene affairs at the ValuJet crash in the Florida Everglades was less controversial.

Francis and Hall were not close, having clashed several times during the TWA 800 investigation. Later there was to be disagreement over Francis's foreign travel budget. Francis enjoyed accepting invitations to visit overseas aviation conferences, but Hall sought to limit the number of trips he could take. Francis, a slow speaking, balding, blue-eyed man, had a relaxed, laid-back demeanor and liked to be briefed on the content of a technical report by its author rather than read it himself. At the board level, this tended to slow things up because "Francis would just sit there grinning, and it was obvious he hadn't read the report," according to one senior staffer who attended many board sessions.

Francis was regarded as undependable by some at the NTSB, especially by Loeb, who insisted that staff members be accompanied by their supervisors whenever Francis asked to see them. Loeb suspected that Francis was still wearing an FAA hat, but Francis saw himself as standing up to Hall's dominance over a weak board. "Hall wasn't terribly open as far as the other board members were concerned," Francis remarked later. "He didn't communicate much with us. And he didn't encourage the staff to communicate with us either. He systematically excluded the board members from a lot of important things that were going on.... Other members of the board were not willing to stand up to him, and he was capable of running the place pretty much as a one man show."

Other relationships at the board level were equally uneasy. Bob Francis didn't get along well with his colleague George Black, and neither did senior staffers like Loeb and Hauteur, who doubted they had Black's full confidence.

Black was appointed by Clinton in February 1996 and was the first highway civil engineer to be appointed to the NTSB board. Although he had impressive road accident investigation experience from his 24-year service as an Atlanta highway engineer, his aviation experience was limited to maintenance duties on bombers and in-flight refueling tankers in the US Air Force during the Vietnam era. Despite having to run to catch up on the progress of major aviation investigations under way at the NTSB when he arrived, Black was diligent about keeping himself briefed on developments.

Though they had little in common, Black and Francis shared the same attitude toward the Flight 427 investigation. They felt that too much was being shoehorned into the report to fit the PCU jam-and-reversal scenario without firm evidence to back it. Francis thought the first draft of the report went too far in ascribing a definite probable cause. Black agreed, and both men were unhappy with a previous suggestion that the entire rudder system be redesigned.

Loeb was in trouble because one of the five board members, John Goglia, could not vote because he had been a party member to the early stages of the Flight 427 investigation prior to his appointment to the board in August 1995. As a USAir mechanic, he had been a prominent member of the International Association of Machinists and Aerospace Workers and served as its flight safety representative for 21 years. Goglia more or less supported the Loeb/ Hauteur line on Flight 427, but because he could not vote, the board was split 50/50, with Jim Hall and John Hammerschmidt supporting the report.

Hammerschmidt had a long association with the NTSB, having been a special assistant to the chairman between 1985 and 1991. Previously, he had worked in the White House as a vice-presidential aide, and was appointed a board member in 1991. Hammerschmidt lacked any prior aviation industry experience, but he did have a private pilot's license. He appeared to have readily accepted the view being advocated by Loeb, Hauteur and their teams. But with a split vote, major concessions needed to be made to Black and Francis.

Hauteur and Loeb had initially hoped to produce their final report by early January 1998, but that was to be the first of many missed deadlines. For one thing, Malcolm Brenner was still turning out human performance reports as late as July. By the fall, Hauteur's team was again revising the report as Dennis Crider continued his work on aircraft performance studies.

In late 1998, Crider decided to make further changes in his reports. He eliminated a 3° rudder movement prior to the supposed moment when the secondary slide jammed. Crider also altered a stabilizer command, based on a revision of the cockpit voice recorder transcript. (The sound of the clacking of the stabilizer wheel was the only way investigators could confirm that it was being operated.) And he discovered an error in Boeing data that predicted the amount of rudder blowdown, which convinced him that pilot pressure on the rudder pedal was not as insignificant as he had at first imagined. In addition, based on further analysis, he decided that the initial wheel command was not as great as he first calculated. In order to fly his simulation through the flight data recorder points, Crider also had to adjust the position of the jam of the secondary slide so that he could match the rate of rudder movement.

To a casual observer, unschooled in the finer points of simulation and kinematics, it could easily appear as if Crider and other investigators were making it up as they went along. It added to Bob Francis's speculation that data was somehow being shoehorned to meet Hauteur and Loeb's PCU jam hypothesis.

Things were dragging on so much during 1998 that three of the parties, Boeing, USAir, and ALPA, decided to put in further written submissions even as they vigorously lobbied the board members.

In August, ALPA said that it stood by its earlier September 1997 submission but wanted to refine its position in the light of new findings, particularly the results of kinematic and simulation studies of the Eastwind Flight 517 incident and new interpretations of human factors findings. The rate of rudder movement arising out of an NTSB simulation study of Flight 517 was exactly the same as would be expected from a secondary slide jam, and ALPA noted a similar finding had emerged from simulation studies of Flight 427 and Flight 585. "It is extremely unlikely that three different pilots in three different B737s, on three different days would use the same rudder rate."

In addition, later human performance studies by Malcolm Brenner, completed after ALPA's earlier submission, reinforced the union's opinion that the accident was precipitated by a rudder jam. Brenner quoted experts like Alfred Belan of Moscow and Scott Meyer of the Naval Aerospace Medical Research Laboratory, who determined that the grunts made by Flight 427 pilot Charles Emmett at the start of the accident indicated significant muscular exertion under a high physical load. The straining sounds began just as the rudder started to go left. "Why would a perfectly fit man need to strain in order to press a perfectly functioning rudder pedal?" ALPA's submission asked rhetorically. Pushing the left rudder pedal in the normal course of events would not produce much strain. But what, ALPA continued, if he were pushing a right pedal which, under the reversal scenario, had come relentlessly up against his foot and defied all attempts to center it?

> ... Under these circumstances, the strength that the first officer likely used while attempting to press on the right rudder pedal would have required muscular exertion, physical straining, increased muscular exertion, high intensity physical activities, straining, great physical effort, strain, high physical loads and struggling. These, of course, are the exact words that the experts used to describe the first officer's speech utterances....

USAir, on the other hand, was dismissive of any attempt to link Emmett's grunts with the operation of any cockpit control. Meyer had suggested that the first officer was straining in a bid to overcome the autopilot. (Large, energetic movements of the control wheel will cause the autopilot to cut out.) "Dr.

Meyer's speculation that noises made by the first officer indicate that he was making certain flight control motions is not based on fact and should be disregarded," wrote the airline.

USAir devoted the remainder of its second submission to robustly challenging points made by Boeing in its first submission. "... a finding that the persistent full left rudder input was made by USAir Flight 427's flight crew must necessarily rest solely on rank speculation which is itself based on anecdotal reports inaccurately and misleadingly interpreted and in many cases taken out of context." On Boeing's suggestion that the pilots had misapplied the rudder after being startled by the wake encounter, USAir was scathing, saying that to imply "... that pilots can be so startled by a wake vortex encounter that they, in effect, forget how to fly is a leap in logic wholly unsupported by the facts." USAir's submission then pointed out that in the six incidents offered by Boeing to highlight the fact that crews get startled, all six crews safely recovered control of their aircraft. The submission also took issue with Boeing for implying that airline pilots were not properly trained:

> ... Boeing's submission states that airlines are now teaching pilots to use aileron and rudder to recover from roll upsets, implying airline pilots have not been taught to do so before. Pilots, including airline pilots, have always been taught, beginning with their very first flight as a student, to use coordinated rudder and aileron in a turn. To imply otherwise is simply misleading and untrue....

USAir also criticized Boeing's use of the Sahara Airlines case, which—if pilot error was to blame—showed "the error was made by a low time pilot with no large jet aircraft experience. This is not an incident from which a valid inference can be drawn about the possible actions of pilots with thousands of hours of airline experience."

Boeing's new submission concentrated on the latest analysis of the Eastwind Flight 517 incident, saying that there was no evidence that it, Flight 585, or Flight 427 had been caused by a rudder PCU malfunction. The most likely explanation for the Flight 517 incident was a yaw damper failure, which subsequently cleared itself, and no additional safety recommendations were needed. The Flight 517 crew's description of the event in interviews afterward, Boeing suggested, was mistaken. Boeing then produced an explanation for their mistaken recollection. "In order to understand the interview data, it is important to understand how one's memory of such an event may be affected by the situation, particularly the element of surprise, fear and sudden response."

In short, Boeing suggested, they had imagined most of it.

apter 20 FINDINGS

At the beginning of 1999, the NTSB was seething with the final negotiations on the Flight 427 report being written and rewritten by Bernie Loeb and Tom Hauteur. Loeb was attending contentious meeting after contentious meeting with the NTSB's managing director, Peter Goelz, who was attempting to steer the drafts into a final report that would make everybody happy. Side meetings were taking place with, among others, board member Bob Francis's special assistant, Denise Daniels, who was calling for extensive redrafting on his behalf.

In order to get board approval for the report, Loeb and Hauteur had to make many concessions. References to unsafe conditions had to be toned down, and major debates with board members were sparked by relatively simple matters of language. For example, one of the recommendations the authors sought was to make the rudder power control system truly redundant—meaning that if something failed, there would be an immediate backup available. Loeb insisted that the word "redundant" be qualified by "truly." Otherwise, if the recommendation were to make the system simply "redundant," Boeing would insist that the dual concentric servo valve *was* redundant, which it had been saying for years. "So we needed a word to qualify it," Loeb would recall, "and my word was *truly* redundant. Then we had some internal politics and someone, I won't say who, decided they liked the word *reliably* redundant, although that was not something I or my staff ever came up with." But "reliably" it was to be.

Also changed were early versions of the report that sought to dictate in engineering terms how the 737 rudder system should be rebuilt. Board member George Black had his own preferred approach that he insisted be included in the report in return for his support. He wanted the FAA to establish a new probe into the 737 rudder—a new "task force"—and come up with a solution based on its findings.

Loeb was furious about this and the slap in the face of his staff that it represented, but he had no choice. Black had made this one of the conditions for his support, and Francis was backing Black. Without their support, Loeb could forget about steering a report through the board.

Loeb was willing, however grudgingly, to make compromises on the language not because he had worries about further technical reviews of the 737 rudder. He had no doubt a review would confirm the NTSB investigators' findings. His concern, instead, was about the impact of a further delay on the business of fixing the 737. It had been more than four years since the crash of Flight 427. He was worried about delaying a redesign for another year and a half or two years while 737s were flying with the potential for another catastrophe. As Loeb later explained:

> We knew what another catastrophe would mean. The FAA would have no choice and the rudder would get fixed pronto. But we also did not want any of the public to come back to the [NTSB] staff and say "*you* delayed this process, *you* knew damn well this thing should have been redesigned." And so we wanted it to be redesigned right away. That was why we did not want to have to put together this task force; we'd been there, done that before with the same sort of people that would be on this task force....

Loeb's uneasy sense of *déjà vu* was the result of the Boeing 737 Critical Design Review team put together by the FAA in 1994 and the panel of outside advisors assembled to review the work of the Systems Group in 1996. The panel had suggested the cold-soak tests, which led to a breakthrough, but NTSB staffers regarded the Critical Design Review as inconsequential. Outside experts brought late into the investigation meant catch-up time and a learning curve, and new views and new theories that had to be given their due, that had to be taken into account. And always, more delays.

Even if he got the report passed, Loeb felt that its findings and recommendations would be compromised from the outset. The board members—the politicians—would provide the FAA all the cover it needed by saying, in effect, they were voting for a report they didn't really believe in. The FAA would appoint an engineering test and evaluation board (ETEB), and the NTSB could do nothing until the ETEB turned in its report, which could take years.

But when he was under pressure for further revisions to the report, with no guarantees that it would be accepted, Loeb had an unexpected stroke of good luck—if one can consider a possible disaster as good luck.

The incident occurred on February 21, 1999, with 117 passengers and crew aboard a MetroJet 737 at cruising altitude, 33,000 feet, on a flight between Orlando and Hartford. The captain noticed the control wheel moving left, although the aircraft did not appear to be rolling or deviating from course. When he disconnected the autopilot, however, he found that the aircraft started rolling to the right, and that the right rudder pedal was fully depressed even though it had not been pushed. He manually applied left aileron and adjusted the engines to balance the aircraft and stop the right roll. Then he pressed the left rudder pedal to center his rudder, but it was jammed. The other pilot subsequently also tried and failed to move the pedal when carrying out a technique then being promoted by Boeing. They switched off the yaw damper. That did not center the pedals, so they carried out a new recovery technique being promoted by ALPA: they disconnected the two main hydraulic systems and reverted to the standby system. This helped clear the jam, but the pilots noticed that the rudder "pulsed" for some time afterward. Alarmed by his failure to understand and, more importantly, manage the balky rudder, the captain ordered an emergency landing at Baltimore-Washington.

Although the NTSB retrieved good information from the flight data recorder by the time the pilots taxied in, the CVR tape, a 30-minute continuous loop, had recorded over a large portion of the cockpit sounds from the incident. Jim Hall used this loss of data to mount a further attack on the FAA's tardiness in making longer-running recorders compulsory in all aircraft. But he refused to delay a decision on the Flight 427 report in order to take the MetroJet incident into consideration. It was not a near-fatal event and certainly not as dramatic as Eastwind Flight 517, but it did have one particularly disturbing feature. The aircraft was sporting one of the new PCUs, which had been designed to prevent uncommanded rudder movements, but the rudder still went to hardover at its blowdown limit.

The MetroJet incident was interesting in other ways as well. First, compared to the more or less linear movement of the Flight 517 and Flight 427 rudders—single, steady changes that sent the planes plunging out of control—the MetroJet rudder moved in two phases. And second, it moved much more slowly than in either of the other two incidents, leading investigators to wonder if the PCU was at fault at all. And in fact, when the PCU was taken to Parker Hannifin, it was found to be in perfect condition, displaying no evidence of any malfunction.

John Cox, ALPA's representative on the Flight 427 Systems Group, followed the MetroJet investigation closely. "With typical rudder reversal the rudder moves at a predetermined rate, between 20° and 23° per second. If you have a problem with the rudder outside of that range, then you have a different phenomenon in play," he insisted. The MetroJet rudder movement was much more leisurely, he said, "in the neighborhood of 2.3° to 3.0° per second."

The MetroJet incident was not an isolated incident. The previous weekend, a United Airlines 737 pilot decided not to take off from Seattle when he detected stiffness in his rudder pedals in a preflight check. The left pedal was sluggish, and the right one would only move after the application of considerable force. When the PCU was removed and taken to United's San Francisco maintenance base, the input crank—the rod that conveys pedal movements to the PCU—was found to be off-center. Dismantling the unit revealed that a spring was improperly installed and was keeping the secondary slide off-center. It turned out that this unit had been used by United mechanics to demonstrate how the new PCUs were to be tested for the cracks that were found in some of them. (The cracks were first discovered in new units straight off the production line at Parker Hannifin.) The mechanic responsible could not recall fully checking the unit afterward.

A grim joke quickly made the rounds that Bernie Loeb had somehow arranged the MetroJet and United incidents to convert any "undecideds" on the Flight 427 report. Loeb later admitted that he could not have planned it better himself. The events rejuvenated his Flight 427 campaign within the NTSB and helped him argue for some items he was still fighting for.

But the FAA and Boeing started an aggressive lobbying campaign to convince anyone who would listen that the Boeing 737 was a perfectly good design that needed no more fixing. Their targets were NTSB staffers and board members, members of the party system—especially USAir and ALPA—and the media. The FAA's director of aircraft certification, Tom McSweeney, who led the roadshow for his agency, was not averse to using Boeing promotional material to make his point. Boeing supplied him with a video presentation showing how the later fixes, like the hydraulic pressure limiter designed to prevent the rudder from swinging too far above 1,000 feet, which had been announced by Al Gore in January 1997, would mean minimal interference with flight in the event of an uncommanded rudder movement. McSweeney was also provided with a servo valve with the outer casing cut away to reveal its inner workings, which he used to demonstrate how, if one slide jammed, the other was capable of carrying on. It was ingenious, he insisted, two parts in one, a device with its own built-in backup. Perhaps conscious that Loeb intended to make a case for the recertification of the 737, McSweeney praised the US system of aircraft certification as the best there is. He also defended Boeing's right to do its own designs, a move to forestall any recommendation from the NTSB that Boeing install a different rudder system, such as a split-panel design. Such interference by a federal agency would, he insisted, stifle initiative.

The incongruous nature of the situation was not lost on Loeb. Here he was, a government employee acting on behalf of the American people, using their tax dollars to make flying safer, and being opposed by another civil servant, also spending taxpayers' money in promoting a point of view favored by a multibillion-dollar corporation.

As Loeb saw it, the game for Boeing was not one of believing they would win in the end but of delaying, which the FAA had been willing to play along with for years simply because rudder-induced disasters were exceptionally rare:

> ... The kind of events we are looking at are rare events, which is why we had so much difficulty investigating them. If it were to happen again, it would be devastating to Boeing but it would also be a significant problem for the FAA, which has unfortunately not stepped up to the plate the way I believe they should have.

Loeb agreed with the widely held belief that the FAA still regarded the promotion of aviation as a prime goal, even though this had been dropped from its charter some years earlier. The agency, Loeb believed, still saw itself in some respects as a promoter:

> ... The notion of the promotion of aviation was something that was deeply embedded in the psyche of the FAA organization and I think that has continued to be a problem. There continues to be a difference of opinion between me and many in the FAA on what their job is right now. They have been aided by the Supreme Court, to some degree, because there was an argument over whether or not they were the guarantors of aviation safety and the court essentially said they were not....

Later, chilling details would emerge of how much the FAA was marching, hand in glove, to industry's tune. When some relatives of Flight 427 victims through legal processes obtained Boeing documents relating to the crossover speed issue, they were amazed to discover that this had been discussed in secret with the FAA in September 1992. Boeing insisted the material under discussion was proprietary and forbade any release of it to the public. The FAA meekly complied and took no further action. Yet, armed with this information, the FAA had the opportunity to review the Boeing 737 flight manual and to decide if pilots needed to be made aware of the hazards of a rudder hardover at slow flying speeds, where the ailerons could be overpowered by the rudder causing an aircraft to crash. Later, FAA test pilots working on the Flight 427 investigation discovered the true nature of the crossover speed phenomenon, not from those 1992 disclosures, but from independent simulator experiments. It can be legitimately argued—as USAir did—that the pilots of both Flight 585 and Flight 427 might have tried alternative, and possibly successful, methods of saving their aircraft had they been aware of the crossover speed issue.

In February 1996, FAA inspectors based in Atlanta recommended the grounding of the budget airline ValuJet following repeated aviation code violations and safety-related incidents, including dozens of emergency landings. Their report was shelved, and no action was taken. Three months later, a ValuJet DC-9 crashed into the Florida Everglades after an in-flight fire, killing all aboard—yet in the wake of that accident, senior FAA officials, including administrator David Hinson, went on TV to defend the airline's safety record. Later it would emerge that ValuJet had 14 times more violations and safety incidents than the average US airline, and the FAA was finally forced to remove its operating licence.

On February 24, 1999, a month before the public board meeting scheduled to deal with the Flight 427 report, Loeb formally handed it over to the board members for their final deliberation. Included was his appeal for the rudder to be redesigned, which he knew at this stage was doomed. Making it "reliably redundant" would have to suffice.

The final board meeting on Flight 427, the so-called "sunshine" meeting, took place March 23 and 24, 1999, at the NTSB's headquarters at L'Enfant Plaza in Washington. Loeb and his staff made a number of well-rehearsed presentations. These included a series of computer-generated video simulations showing how the NTSB investigators believed the doomed plane turned, then inverted and dived into the ground. Other simulations were prepared of the Colorado Springs accident and the Flight 517 incident. Of all the presentations, Malcolm Brenner's human factors presentation was one of the most persuasive in putting forth the rudder-reversal theory. A natural storyteller, Brenner convincingly evoked the sounds on the cockpit voice recorder tape as a human bridge between the dry facts of Crider's simulation and the obscure and hard-to-grasp details in Greg Phillips's account of the misbehaving PCU.

Brenner had the audience spellbound as he brought to life the last moments of the 427 pilots as they struggled with their uncontrollable aircraft.

He explained how the grunts and strained breathing of First Officer Emmett could be matched up with moments when the rudder would have moved left to hardover and the right rudder pedal would have come up forcefully against his foot.

But first, Brenner demolished Boeing's theory that the pilots were startled by the wake vortex encounter and panicked into mishandling the controls.

> ... When rudder pedal was put in the airplane was still in normal flight and they were dealing with a wake turbulence event. We looked through the literature to see if we could find precedents which suggested that this would have happened. There were three crashes related to wake turbulence but all three were in the late sixties or early seventies and were under the old separation rules [when aircraft on landing approaches followed each other more closely]. None had occurred at the altitude of Flight 427....

Brenner then described how he had searched through the Aviation Safety Reporting System, a confidential database operated by NASA for pilots to anonymously report hazardous situations without triggering a formal investigation. He found 100 reports related to wake encounters.

> ... Most were at lower altitudes than flight 427 but they all recovered safely. Wake turbulence just isn't a severe event. Our flight tests show that it's difficult to stay inside a vortex, you get thrown out again, and besides, the autopilot was well able to handle it. These pilots had encountered wake turbulence before. Why should this time be any different? The first officer makes a clear progression of actions. He jumped on the controls, he broke off speaking to the captain and he said "Zuh," a clear expression of surprise. In less than a second he made a hard right input on the controls to bring the airplane back. As the airplane began responding he backed off on the controls, an understandable and very alert pilot reaction. At this point we believe he entered some right rudder to correct the yaw that was involved and experienced a rudder reversal. Suddenly he had a new situation which he could feel on the sole of his foot, the rudder pedal would not go down. That's when the grunting begins. I believe there he made an effort to break the jam on the pedal. But at the same time he's having trouble with the main control, the wheel. At this point, both our model, and the pilot input model [Boeing's] show that he was making very aggressive motions on the wheel to level the airplane. Then he's discovering something else for the first time in his career. The wheel does not have the authority that he expected to bring that airplane

> back to level flight. At this point he backs off a little on the controls and disconnects the autopilot. This is a reasonable action and a sign that he realizes that things are becoming more serious, he's wondering if the autopilot has anything to do with it. He's quiet for a little over 2 seconds, and then he swears. He says it very quietly, he's not straining at this stage, he's not fighting with the pedal although he's putting in a lot of force. To me it sounds as if this is more than wake turbulence, there's something else going on here. The captain has been coaching him up to this point. He's been saying "Hang on, hang on, hang on." At this point the captain becomes quiet, the airplane is now leaving the attitude of normal flight. A few seconds later the captain says "What the hell is this?" During this period the first officer is pulling back on the yoke (wheel), the airplane is getting into a more severe roll, the ground is coming up. Then the stickshaker activates and draws attention to the fact that they are about to stall. The captain makes an emergency call to air traffic control, with just a few seconds to impact with both pilots pulling back on the controls, they're trying to raise the nose of the airplane, they're trying to deal with the situation but they couldn't figure it out. They didn't know what they had.

After a day of presentations by Brenner and others on the staff and pre-arranged questions from board members, Jim Hall read out the 34 findings they had agreed upon and, finally, the probable cause of the accident.

In essence, Flight 427 had been a perfectly good, well-maintained airplane, properly operated, on which nothing had broken, exploded, or fallen off prior to the accident. Apart from an uncommanded rudder movement, there had been no engine or control malfunctions. The crew reacted properly and immediately according to their training, and did not apply a full left rudder, nor hold it until impact.

Finding number 11 was the first nail in the coffin of Boeing's effort to achieve a different result. It read:

> ... Analysis of the cockpit voice recorder, National Transportation Safety Board computer simulation, and human performance data (including operational factors) from the USAir flight 427 accident shows that they are consistent with a rudder reversal most likely caused by a jam of the main rudder power control unit servo valve secondary slide to the servo valve housing offset from its neutral position and overtravel of the primary slide....

It got worse for Boeing. Finding number 13 read:

... The flight crew of USAir flight 427 could not be expected to have assessed the flight control problem and then devised and executed the appropriate recovery procedure for a rudder reversal under the circumstances of the flight....

And worse again. Subsequent findings stated that it was unlikely that Flight 585 was caused by a mountain rotor but was consistent with the same causes just read out for Flight 427. The Eastwind Flight 517 incident was also blamed on a servo valve jam. (The cause of the jam, in turn, was not specified: it might have been some combination of a tight fit, thermal effects, contaminated fluid, or other, unknown factors.)

Then there followed some good news, but mostly more bad news for Boeing. When completed, the rudder redesign changes then under way should prevent the same thing from happening again. But the changes would not eliminate *other* failure modes that could lead to a loss of control. The PCU dual concentric servo valve on the 737, as originally certified by the FAA, was not "reliably redundant." A reliably redundant device was needed, despite "significant improvements" made in the PCU's design. In addition, pilots needed training in avoiding or recovering from an uncommanded rudder hardover. An airplane flying under the crossover speed "presents an unacceptable hazard." Three items relating to the flight data recorder rounded out the findings. The final finding, which stung the FAA and Boeing, read as follows:

... The Federal Aviation Administration's failure to require timely and aggressive action regarding enhanced flight data recorder recording capabilities, especially on Boeing 737 airplanes, has significantly hampered investigators in the prompt identification of potentially critical safety-of-flight conditions and in the development of recommendations to prevent future catastrophic accidents....

The probable cause statement was short. But it did not, as had been feared, list failures by Boeing or the FAA as contributors to the accident. It read:

... The National Transportation Safety Board determines that the probable cause of the USAir flight 427 accident was a loss of control of the airplane resulting from the movement of the rudder surface to its blowdown limit. The rudder surface most likely deflected in a direction opposite to that commanded by the pilots as a result of a jam of the main rudder power control unit servo valve secondary slide to the servo valve housing offset from its neutral position and overtravel of the primary slide....

The technical language of the probable cause statement was just that—technical language, and not a statement of blame, or a statement of failure. Similarly, there were ten Safety Recommendations addressed to the FAA that focused not on opportunities missed or responsibilities overlooked, but on just what needed to be done going forward. First, that all existing and future 737s have a reliably redundant rudder system. Second, that the FAA convene an engineering test and evaluation board (ETEB) to inquire into the 737 rudder. The third and fourth recommendations were aimed at passenger aircraft generally and called for them to have a reliably redundant rudder system as well as be capable of a continued safe flight and landing in the event of a jam of any control surface. There were three recommendations aimed at improving the training of pilots to meet rudder jams, plus two recommendations to settle the issue of the crossover speed hazard, an issue also dealt with under training. Finally, and not unexpectedly, a recommendation geared to dramatically increase in a two-and-a-half-year period the numbers of flight parameters measured by flight data recorders on all US airliners.

Loeb was disappointed that he had not achieved more, but could console himself with the fact that no air accident investigation report had ever before been subject to so much determined opposition and lobbying by the FAA and a plane maker. Even before the report was published, Boeing's anger at the performance of its lobbyists led to the sacking of John Purvis, Boeing's director of flight safety, and the removal of chief 737 engineer Jean McGrew from the investigation. Soon afterward, bitter at how he had been treated, McGrew took early retirement.

Not surprisingly, the NTSB failed to maintain a united front. After the meeting, Chairman Hall pointedly told reporters he flew a 737 to Chattanooga regularly for the weekend and felt perfectly safe. George Black was overheard saying that he thought the findings were so heavily based on circumstantial evidence that he almost voted against the report. Some time later, Bob Francis admitted that this was the "least comfortable vote" he had cast as a board member, despite getting his way in amending the report.

Shortly after the board meeting, the FAA started bad-mouthing the findings, even though Boeing initially appeared to welcome them. The FAA's response was to point out that the cause of the rudder movement was still unproven.

"Data do not conclusively support any specific scenario," read one of the headings in a PowerPoint presentation developed at Tom McSweeney's direction. There was no evidence of a jam caused by a metal chip in the valve of the

427 PCU, the presentation maintained, and a thermal jam had been created only in the laboratory and therefore could be ruled out. The FAA and Boeing had already addressed and fixed all possible scenarios for the accident, it added, and the issue of possible rudder reversal had been addressed by the 1997 redesign of the PCU, in which the possibility for overtravel of the primary slide had been eliminated.

However, in May the FAA said it was taking action on the flight data recorder recommendations. The FAA's new head, Jane Garvey, said it would heed the NTSB's recommendation to add rudder and rudder pedal positions to flight data recorders in addition to upgraded requirements announced two years earlier. Cockpit voice recorders were also to be improved, she added, with a two-hour capacity replacing the older 30-minute recordings. All flight recorders would also be equipped with independent power supplies to keep them working for 10 minutes after an aircraft's power failed. In the same month, the FAA announced the composition of the ETEB task force and set it to work. It seemed that the other recommendations would have to wait until the ETEB report was published.

And Boeing itself issued an official statement on the same day as the final NTSB meeting:

> Boeing is beginning an immediate review and developing action plans to address all of the conclusions and recommendations announced today by the NTSB that pertain to Boeing.
>
> Boeing welcomes and will fully support an independent review and test program for the 737 rudder control system. If there is a way to make the 737 safer, we will take the necessary steps to accomplish that goal.
>
> Boeing also agrees with the board that upset recovery training for pilots is a valuable safety enhancement.
>
> Also today the NTSB released a probable cause for the Pittsburgh accident after a four-and-a-half year investigation and concluded that changes currently being made to older 737s eliminate the possibility of a rudder reversal (the event cited as the probable cause). The NTSB also concluded that further testing and studies are needed, and Boeing is committed to working with the board, the FAA, and the airlines to ensure that everything is done that can be done to ensure safety.

Boeing then claimed it had taken numerous actions to address the issues raised in the Flight 427 investigation: it had improved its design of the 737 rudder system and the yaw damper system; it had added a pressure limiter to reduce the effect of the rudder and help the crew better manage incidents; it had added maintenance checks and in-flight procedures to help crews respond in the extremely unlikely event of a rudder malfunction; it had worked with the rest of the industry to develop training to prepare pilots for incidents, regardless of cause.

> The 737 is a safe airplane. In more than 30 years of service, the 737 has carried more than 5 billion passengers on 77 million flights. More than one million people fly every day on 737s; somewhere in the world, there is a 737 taking off every six seconds.
>
> Boeing's commitment to the 737—and all of our other airplanes—remains the same: when the facts show us an opportunity to make it even safer, we will. It is our job to provide airlines and their passengers with safe, reliable airplanes.

But three months later, Boeing spokespeople told reporters that they were confused by the NTSB Safety Recommendations. "We don't quite know what 'reliably redundant' means," said Erik Dixon, a Boeing public affairs spokesman. "That's a new term, we need to understand what that means. It's not a known term in the industry. Once we've figured out what it means we need to know what to do to bring it about."

By this stage, Boeing had developed amnesia about certain aspects of the investigation. When asked if he thought that Boeing had been particularly unlucky in its experience of rudder and yaw damper problems, Dixon replied, "I have no information on that." How did Boeing feel about USAir's allegation in its submission that a contributory cause of the accident was Boeing's failure to inform them of the crossover speed hazard? "I'm not familiar with that material," was his reply. Or about the NTSB's rejection of the rotor theory for the Flight 585 crash? "I don't have the specifics on that one. Wouldn't be proper for me to speculate on that."

Yet Dixon could speak knowledgeably about other aspects of the investigation, such as why Boeing put so much emphasis on the pilots and their reactions when startled.

> ... We were open to any reasonable possibility which included rudder problems but we also included the possibility of pilot error. Those two theories were the strongest and since neither could be proven we wanted to stick with the facts. We wanted to keep an open mind and people may have misunderstood our open-mindedness as being trying to go down one path or another....

Dixon said that Boeing was continuing to "review" some of the other recommendations, especially the two that included the phrase "reliably redundant" and the third one, which required an airplane to be capable of safe flight after a control surface jam. "The study is long term in nature to figure out what to do there."

There was more confusion over recommendation number 5, which wanted improved procedures for recovering from a jammed rudder that "do not rely on the pilots' ability to center the rudder pedals as an indication that the rudder malfunction has been addressed...." Boeing stated:

> ... We've been working with the FAA and the NTSB to understand what is behind this recommendation. This is a procedural type issue—we are trying to understand what they mean by this....

Chapter 21 EPILOGUE

In a corner of a hangar at the sprawling Boeing plant in Renton, just outside Seattle, stood something you don't see every day. A towering tail fin, salvaged from a scrapped Boeing 737, was held firmly upright in a steel frame. It looked like a weird totem erected to aviation technology. The rudder was still attached but part of the fin was cut away to reveal a cavity, within which lay the complex mechanism that moves the rudder. The cavity was festooned with wires and tubes attached to sensors and gauges within it and to pumps and computers outside. Nearby on the hangar floor was a pilot's seat with a set of rudder pedals connected by cables to the tail fin.

On the apron outside the hangar, engineers had set up a maintenance scaffold beside the tail fin of a parked jetliner. This airplane, a 737-200, belonged to Purdue University's aviation technology program, and the university's name was painted on the fuselage. The engineers were busy wiring up sensors to the workings of the plane's rudder system in a careful duplication of the work that had been done inside the hanger. Inside the plane, the first third of the seats had been removed from the cabin and replaced by banks of computers and instruments. The FAA's Engineering Test and Evaluation Board (ETEB) was at work.

Formed in May 1999, just two months after the NTSB wound up its investigation into Flight 427, the ETEB had gone to work almost immediately, first reviewing the thousands upon thousands of pages of research documentation produced by the NTSB, by the FAA's earlier Critical Design Review team, and by Boeing itself. Then it set about acquiring an airplane and parts of airplanes to find out what made the 737's rudder work—or not work.

The ETEB's 24 members were drawn mostly from FAA and Boeing engineers, but with the strict proviso that they be leaders in their fields and not have worked at all on the 737 rudder system, either at the design, development, or certification stage. Only one of them had any connection with the NTSB, and he had no significant involvement with the Flight 585 or Flight 427 investigations.

The team leader, John McGraw, was manager of the Airplane and Flightcrew Interface branch of the FAA's Transport Airplane Directorate in Renton. He was a former military flight test engineer who had worked on both army and navy projects, including testing the US Army's Kiowa Warrior helicopter. Technical leader of the ETEB was Boeing's Dagfinn Gangsaas, who had led flight control development for Boeing's Joint Strike Fighter program. More recently, he had been chief engineer for the Darkstar unmanned military aircraft program and led the redesign of the remote-controlled plane after its prototype crashed in 1996. The FAA's Jim Treacy, a former Lockheed engineer in charge of systems safety on the ETEB, was an internationally renowned expert on electronic navigation and control of aircraft, with a sideline in the control of random failures. David Wineman came from NASA with a background in hydraulic troubleshooting, including work on a program to cure hydraulic failures in the F-14 fighter. There was even a Russian on the team: Sergey Boris was deputy director of the Flight Research Engineering Center in Moscow and a systems expert.

The projects listed on the résumés of the other experts read like an honor roll of recent US aeronautical history, but there was one exception. Davor Hrovat, a Ford Motor Company control engineer who held numerous patents in vehicle power-train controls and was the winner of several prestigious national engineering awards, was chosen because of his experience at the leading edge of automotive control and braking systems, which would give the aeronautical engineers a different perspective on their work with complex hydraulics systems.

The tail fin in the Renton hangar was hooked up to a series of hydraulic pumps, just as in a complete airplane. Computers recorded pressures, temperatures, input forces, and output forces. If there was a useful probe or a gauge that could be fitted to the system, it was attached. Video cameras recorded what could not be otherwise measured. The airplane outside was set up the same way. It was in perfect flying condition, but like most elderly US 737s, it lacked adequate flight recorders. That was soon rectified by Boeing wiring in a portable system.

Having access to a working airplane for several months was a luxury enjoyed by neither the Critical Design Review team nor the NTSB. Indeed, to acquire and experiment with a tail fin alone was something Greg Phillips could only dream of. In addition, the ETEB had a PCU servo valve that showed its inner workings. Fitted with special probes, it revealed the positions of the two slides relative to each other and to the valve as a whole. These probes could tell immediately if the valve was working normally or if it was jammed, and if it was jammed, what force it was applying to the rudder actuator. This servo valve was installed on the tail fin in the hangar. Remarkably, this was the first time there had been such a complex simulation of any workings of the 737 since the 1960s, when Boeing had rigged up the so-called "Iron Bird," a non-flying laboratory where all the working parts of the 737 were mounted, powered up, and exhaustively tested before being tried out on a prototype aircraft.

The ETEB had a third weapon in its armory: access to the Boeing M-Cab flight simulator, where they could try out scenarios before taking to the air. They also planned to use it to assess how pilots react to rudder emergencies.

The ETEB's probe was to be conducted without regard to existing aviation regulations or the low probability of an event. In other words, FAA regulations allowing a condition to exist was no bar to the ETEB's declaring it a hazard. And the fact that a hazard might be so obscure or statistically unlikely to happen, like a bird striking the nose of the aircraft and wedging its carcass underneath a rudder pedal, was of no consequence. If it could happen at all, it was of interest.

Sixty-six separate tests of different rudder and yaw damper system conditions and failures were planned. The rudder controls in the tail fin in the hangar were encased in an insulated environmental control chamber in which icing and heating could be produced. Flights in Purdue's 737 were used to evaluate both older and post-1997 PCUs, measuring the effect of variables such as temperature and wake vortices. The wake vortex tests were conducted behind the FAA's Boeing 727. Flights were also conducted to calibrate the simulator, monitor the dynamic performance of rudder components, and experience yaw damper hardovers. A total of 11.5 hours of flight tests took place between December 1999 and March 2000. Lab tests were carried out on a variety of PCUs and included a repeat of the earlier cold-soak tests. New tests were devised to determine the effect of vibration on the unit, something that had not been assessed before.

In July 2000, the ETEB issued a bulky 950-page report, more or less on schedule. It was a stunning eye-opener for both Boeing and the FAA. The report listed 46 failures and jams in the 737 rudder system that could have catastrophic effects. The standby system, the report concluded, was not truly independent of the main rudder system. Some faults were caused by a newly discovered threat, icing. Maintenance procedures were not catching some failures that had no immediate effect but could later lead to accidents. Most pilots did not understand rudder malfunctions and were poorly trained to deal with them.

The old PCUs, those unmodified by the 1997 directive, had serious problems. Fifteen single failures and jams were found that could have catastrophic effects on landing and taking off, plus 12 other potential hazards. The modified PCUs with a pressure limiter were almost as dangerous: they could suffer 14 single catastrophic failures and jams in addition to 12 potential hazards. The modified and unmodified PCUs had dozens of other failures and combinations of failures that could also prove hazardous, although not necessarily catastrophic.

And in the course of studying the plane's vulnerabilities, the investigators went well beyond anomalies that already existed in the tail. Although it performed no specific tests to prove or disprove the contention, the ETEB said that the 737's hydraulics and cable systems were vulnerable to a collapse of the cabin floor due to explosive decompression, or to a large bird (4 pounds or more) striking the nose cone or the curved fiberglass fairing where the front of the tail fin merged with the fuselage. In either case, rudder control could be lost. The engineers were also concerned about the effect of a destructive engine failure that would eject fragments and damage main and standby hydraulic lines or rudder cables. Without an independent rudder system to compensate, the asymmetrical thrust of a single remaining engine could drive the aircraft into the ground.

The ETEB was also unhappy with the emergency procedure Boeing prescribed for a jammed or restricted rudder, which it declared confusing and difficult to use after trying it in the simulator. The procedure first called for disconnecting the autopilot. Then the pilots were expected to free the jam by applying force to the pedals, working together if necessary. If they could center the pedals, the procedure called for a normal descent and landing. If they could not center the pedals, the procedure called for switching to the standby rudder hydraulic system. One problem the ETEB found was that the considerable force applied to the pedals often centered them by stretching the cables; this was deceiving because the rudder jam would persist and make the aircraft dangerous to land. Yet the next step in the procedure was for the volunteer test pilots to "land" the aircraft simulator. "They were in a dilemma, they knew the procedure was not working, but they were reluctant to deviate from a formal procedure," the report stated.

Letting different pilots attempt to deal with jams, hardovers, and reversals in the Boeing simulator demonstrated that there was confusion in the ranks. It emerged that virtually all pilots were inadequately trained in the procedures to use. "None of the crews could execute the checklist the first time, and most crews could not correctly execute the checklist after several attempts," reported the ETEB.

The ETEB was alarmed at pilots' failures to properly diagnose the condition of the airplane after a jam or a hardover and to decide if it was safe to attempt a landing. On nine occasions, the flight crew in the simulator completed the jammed or restricted rudder emergency procedure but still ended up with an aircraft unsuitable for a landing.

In addition, pilots were confused about the extent to which they could press the rudder pedals. The 737 was almost unique among airplanes in having a restricted pedal travel at high airspeeds, about 1.5 inches, while at low airspeeds the pedal travel was more than 4 inches. Yet when questioned about this, most pilots were unaware of the differences. The ETEB feared that pilots might misinterpret the low travel at high speeds as evidence of a jammed or restricted rudder. Simulator training was of limited use because many simulators did not have adequate software to replicate a jam, especially the large pedal forces involved.

Pilot training was found to be woefully inadequate for consistent recovery from large roll and yaw movements at low altitudes. In the simulator, most pilots pulled back on the control column before the airplane achieved a level position, which often led to a crash. "Any sustained aft column input during the upset recovery resulted in ground impact," the ETEB report tersely stated. The industry-wide training on stalls at the time emphasized remaining at the same altitude, whereas the best recovery technique was to trade some altitude for speed. Overall, simulations showed that rudder hardovers at low altitudes were likely to be catastrophic. When pilots were unexpectedly given a simulated rudder hardover at 400 feet at an airspeed lower than the crossover speed, 40 percent crashed, and the others suffered heading changes of more than 90°, which could lead to a collision with tall buildings or high terrain. Contributing to the difficulties was the fact that the pressure limiter does not function at a low altitude, only coming into play above 1,000 feet.

More problems were observed during the approach to landing. Out of 11 crews given a simulated failure, only 6 got the procedure right. One crew drove off the side of the "runway" at high speed, another "landed" with a powered-up standby rudder and no power in their ailerons or elevators, although they did make a successful landing. Three other crews ended up with the A and B hydraulic systems still applied to the main PCU and the rudder deflected to its blowdown limit. They incorrectly concluded the rudder could be controlled and attempted a normal landing.

And sometimes, the ETEB discovered, shutting off the yaw damper, as called for in some of the recommended recovery techniques in order to stop rudder oscillations, resulted in a full rudder hardover. "Some flight crews and maintenance personnel were not aware that rudder kicks are indications of a potentially serious flight control problem."

Interestingly enough, there was a degree of correlation with Boeing's earlier contention that pilots could apply the wrong controls under pressure. In one simulation of a left rudder hardover, a first officer hesitated for about 3 seconds, and when the bank angle reached 30°, responded by turning the wheel in the direction of the bank (i. e., the wrong way) and by pulling the column almost fully back. The airplane "crashed" a few seconds later. In one series of simulations of landing in a crosswind without rudder, one crew attempted to keep their airplane centered on the runway using differential reverse thrust, but applied a greater thrust to the wrong engine and went off the side of the runway.

Not all potential problems resulted in jammed pedals. The ETEB team investigated one scenario in ground tests in which the external linkage of the PCU was artificially jammed. If the PCU linkage was then disconnected to simulate a break, the rudder deflected to the right and went to the blowdown limit. The rudder pedals seemed to function normally, but the pilots had absolutely no control over the rudder. Only 4 out of 13 crews facing this condition in the simulator were able to correct the problem using the standby system to center the rudder. In some cases, the pilots continued to apply force to the rudder pedals, even though this was having no effect because of the broken linkage. All the pilots eventually managed to get the simulator into a position for a successful landing, although they curiously rated this condition as less severe a test than a more straightforward jammed PCU, where the pedals would jam. "This is likely due to the fact that the pedals feel normal, even thought they were disconnected from the rudder, and had no effect on rudder position," the investigators decided. The lack of a clear indication of rudder position from the cockpit was judged a serious deficiency.

Other failure modes were equally confusing for the test pilots, all 20 of whom were volunteers drawn from four airlines. In some simulations they lost between 3,000 and 5,000 feet in altitude before regaining control. In a real landing approach, this could be catastrophic.

The presence of the pressure limiter, which reduces hydraulic pressure and therefore the severity of a jam, was found to be a mixed blessing. It resulted in a smaller and more easily controlled rudder hardover in the case of an aircraft climbing or cruising above 1,000 feet, but could contribute to an accident in the case of an aircraft coming in for a landing whose pilots were still attempting to sort out a jam. Once the aircraft descended through the 700-foot level, the hardover suddenly became especially severe and added to the pilot workload.

Icing tests on the tail fin were illuminating. In some cases it had no effect; in others, ice accumulated around linkages or stops on the main rudder PCU and caused the rudder to go hardover when the yaw damper was activated. Sometimes strong rudder pedal force (up to 500 pounds) would break up the ice jam. On other occasions, hard compacted ice was impossible to break with pedal pressure. Ice could easily form during flight, the ETEB team decided, from a variety of factors including condensation or frost from naturally occurring moisture in the atmosphere and leaks of water into the tail fin. One scenario, where ice accumulated in the gap between the body of the PCU and the input crank, produced a yaw damper hardover very similar to that experienced in the February 1999 MetroJet incident. Icing was also found to produce a pulsing rudder pedal, which the MetroJet pilots said they experienced. Such pedal movements should always be interpreted as indicators of a serious rudder malfunction, the ETEB team warned.

The standby rudder PCUs had mixed results in the icing tests. On the older 737s (100 to 500 series), they were immune to icing interference. But on the later aircraft, from the 600 series onward (known as Next Generation), standby units had the potential to suffer malfunctions.

For the second time, a major study of the Boeing 737 concluded that there was something seriously wrong with its rudder and called for prompt modifications. More urgently, the ETEB advocated immediate changes to the emergency recovery procedures to make it simpler and easier for pilots to be trained. Out of 24 major recommendations in the report, 14 related to the rudder system and its operation. The first read as follows:

> Modify the Boeing 737 rudder control system to ensure that:
>
> - no single failure or single jam of the rudder control system will cause uncommanded motion of the rudder surface that results in a Class I failure effect,
> - no combination of failures or jams will result in a Class I failure effect, except for those combinations that are shown to be extremely improbable, and
> - no probable single failure or jam will have an effect worse than Class IV.

A Class I failure is defined as catastrophic, a condition that prevents continued safe flight and landing. A Class IV failure, on the other hand, is defined as minor, offering only slight reductions in safety margins, a slight increase in crew workload, and some inconvenience to passengers.

Another design recommendation called for a mechanism to independently shut off the hydraulic power to the main PCU without affecting the rest of the hydraulic system. Hitherto, the only way of completely shutting off the main PCU was to shut off both the A and B hydraulic systems, leaving the aircraft dependent on the reduced power of the standby hydraulic system, which did not provide hydraulic power to the ailerons or the elevators.

Four recommendations were aimed at improving maintenance operations and increasing the numbers of checks that should be performed. Three recommendations called for further research into the icing problem, plus the redesign of the main PCU to prevent ice accumulation. The ETEB noted that the Next Generation series of 737s was fitted with a strengthened forward pressure bulkhead and recommended that the older designs also be fitted with stronger bulkheads to give greater protection against bird strikes. There was also a call for further research into the effect on the rudder system of a collapse of the cabin floor, or of a sudden decompression of the aft pressure bulkhead that results in the temporary pressurization of the tail fin cavity. The effect of a collapsing floor had been covered by a new FAA regulation in 1980, but it did not apply to the 737 because of grandfathering. The ETEB also called for alerting foreign operators of the 737 of the need to replace their main rudder PCUs with the new servo valve as stipulated in the FAA's 1997 directive.

Procedures for recovering from failures were covered by five recommendations. One recommendation called for the development of a simplified jammed or restricted rudder emergency procedure that would be clear and unambiguous and include a check on airplane controllability. Another recommended including all aspects of operating the rudder control system in initial pilot training and in recurrent training, including simulator training. ETEB research had discovered that pilots did not understand how their rudder systems worked. They were also unclear about the effect of a rudder blowdown on pedal travel, the adverse effects of applying aft column during a recovery, and even the concept of crossover speed. Pilots needed to be reminded that pedal position was not a reliable indicator of rudder position, and that pulsing of the rudder pedals could be a harbinger of worse problems. In this connection, the ETEB called for a rudder position indicator to be installed in the cockpits of all 737 aircraft.

Like the NTSB before it, the ETEB returned to the vexing question of flight data recorders and called for the addition of more parameters, such as yaw damper commands, rudder pedal force applied by each pilot, and even the position of the secondary slide in the servo valve. It also recommended giving consideration to the incorporation of additional equipment, such as video cameras, to monitor critical system functions.

The ETEB signed off on its final report on July 20, 2000, and formally handed it over to the FAA, which digested its contents over the next two months. On

September 13, 2000, the FAA announced that it would require Boeing to redesign the rudder for all models of the 737, of which there were 3,400 in worldwide service at the time. In a move then estimated to cost in excess of $200 million, Boeing would be required to provide 737s in the US fleet with a sister servo valve mounted in tandem with the existing servo valve on the main PCU. In the case of a jam of one PCU, the other would be able to overpower it and break out of the jam. The new pilot training and operational procedures called for by the ETEB were to be implemented, as well as additional maintenance checks. But there was no mention of the rudder position indicator called for by the ETEB or of a switch to shut off the main PCU in an emergency.

The remarks made by both Boeing and the FAA were spun into endorsements of the safety record of the 737. Boeing denied it was "fixing a safety problem" but rather was involved in a program of "enhancements" of the aircraft. "The enhancements we're announcing today should be taken in context," read a statement issued on September 14 on behalf of Carolyn Corvi, vice president and general manager of the 737 program:

> ... The 737 family has been, and continues to be, among the safest of all jetliners; in fact its safety record is twice as good as the average for the world's commercial jet fleet. But we believe even this airplane can be enhanced. So we are simplifying flight crew procedures, increasing maintenance oversight, and modifying the rudder control system. We think these enhancements will improve an airplane that already has proven itself in more than 100 million hours of flight....

"Any data shows the 737 is one of the safest airplanes in the world today," echoed John Hickey, manager of the Renton-based Transport Airplane Directorate of the FAA. NTSB chairman Jim Hall was less sanguine in his public statement on the report:

> ... While we are very concerned that some ETEB recommendations will not be adopted—particularly an independent switch to stop the hydraulic flow to the rudder and a rudder position indicator in 737 cockpits—we are pleased that both the FAA and Boeing Aircraft Company agree that there is a need for a redesign to the rudder actuator system. However, before the Board can determine if this will satisfy the goal of our recommendations, we will need to evaluate in detail the proposed design. I hope this redesign and retrofit can be accomplished expeditiously so that the major recommendation of our accident report last year will be realized—a reliably redundant rudder system for Boeing 737s....

The September 2000 FAA announcement said that it would issue a formal Airworthiness Directive (AD) the following summer, allowing Boeing time to design the parts. However, the Notice of Proposed Rule Making process did not commence until mid-November 2001, when the FAA formally published a notice in the *Federal Register*. The soothing language of its prior announcements was dropped, and the notice pulled no punches in declaring the present rudder system unsafe:

> This proposal is prompted by FAA determinations that the existing system design architecture is unsafe due to inherent failure modes, including single-jam modes and certain latent failures or jams, which, when combined with a second failure or jam, could cause an uncommanded rudder hardover event and consequent loss of control of the airplane. Additionally, the current rudder operational procedure is not effective throughout the entire flight envelope. The actions specified by the proposed AD are intended to prevent the identified unsafe condition.

But the notice said little about what precise steps Boeing should take, although it stated clearly that additional controls would be designed to provide redundancy. There would not be, as originally forecast by some commentators, a duplicate PCU, but duplicate servo valves that would receive their commands via separate systems of rods and levers. There would also be a system to alert the crew to rudder system failure, although how this was to be achieved was not spelled out. What was spelled out was the cost, thanks to legislation that requires federal regulators to assess the economic impact of their decisions, and this had inflated dramatically. The original estimate of the previous September of about $200 million to fix the US fleet had soared to $364 million, or $182,000 per airplane, of which $140,000 would be for components.

Boeing was not expected to complete its design until the summer of 2002, when flight-testing would take place prior to certification of the new system by the FAA. According to the notice, operators would have five years to complete installation of the new components.

In one sense, Bernie Loeb had been right to oppose the recommendation that led to the ETEB. It bought lots of time for Boeing and the FAA. If his original recommendations had been accepted by the NTSB board and the FAA, the redesign work could have started as early as the summer of 1999, with installation perhaps beginning in 2000. Since the original NTSB recommendations, numbers in the 737 fleet have soared, and now it will be at least 2007 before all 2,000 aircraft in the US fleet are re-equipped. Nobody can predict when another 2,000 or so aircraft in countries outside of the US will be modified.

A cynical review of this process suggests that the longer it takes, the less it will cost Boeing, which is expected to pay for the new parts. Certainly, when Boeing, aided by the FAA, was lobbying hard against the measures being promoted by the NTSB two years earlier, it could be seen that new noise abatement rules being implemented throughout the industrialized world had sounded the death knell of the older 737-100 and 737-200 models, of which more than 1,100 were built—more than one quarter of all 737s. Depending on where they are operated, many of these aircraft face expensive modification to reduce engine noise, or else they will be sold to an airline where noise is not an issue, probably in the Third World. Inevitably, the Third World market will become saturated with cheap secondhand 737s from the US and Europe. Then these older aircraft, which need considerable amounts of preventative maintenance in addition to noise abatement modifications, are likely to be retired.

Nobody could have predicted the tragic events of September 11, 2001, and their economic impact. Seasoned analysts suggest that the main effect on the aviation industry was to steepen a downturn in air traffic that had already begun. One result was many of these older aircraft being parked in giant airplane boneyards in the Nevada and Arizona deserts, sooner rather than later. Most of them will never fly again, a savings for Boeing of more than $140,000 per plane that does not have to have its rudder system fixed. But in a larger view of the turndown, this may be cold comfort.

The NTSB's March 1999 report on the crash of Flight 427 was controversial even before it was written. The staff, without a smoking gun, could not even rely on the support of the board itself. To this day, some senior Boeing personnel privately insist that the crash of Flight 427 was caused by the crew and not by any fault in the airplane. The ETEB report was a tremendous vindication of the NTSB's stance. It was also a serious embarrassment for FAA executives who had declared that the 737 was a perfectly good airplane that needed no further attention. (For a year after its completion, the FAA refused to publish the ETEB report, and the author had to make a Freedom of Information Act request before the FAA agreed to make it generally available.) But mostly the report was a tremendous vindication for the NTSB investigators, especially Greg Phillips, who had long warned of serious problems in the 737 rudder system and battled for years against official indifference at the FAA. He had first advocated major changes to the 737 rudder in a December 1994 memo that was rejected by his supervisors. It will be at least the summer of 2007 before those changes are finally completed—a wait of almost 13 long years. Yet the original premise of the Flight 427 investigation by NTSB staff was so tenuous that without the validation of the ETEB report, the FAA could have continued to dismiss the NTSB's findings.

And the 737 would have continued to fly with a hidden sting in its tail.

INDEX